International Association of Geodesy Symposia

Ivan I. Mueller, Series Editor

International Association of Geodesy Symposia

Ivan I. Mueller, Series Editor

Symposium 101: Global and Regional Geodynamics
Symposium 102: Global Positioning System: An Overview
Symposium 103: Gravity, Gradiometry, and Gravimetry
Symposium 104: Sea Surface Topography and the Geoid
Symposium 105: Earth Rotation and Coordinate Reference Frames

Sea Surface Topography and the Geoid

Symposium No. 104

Edinburgh, Scotland, August 10–11, 1989

Convened and Edited by
Hans Sünkel
Trevor Baker

Springer-Verlag
New York Berlin Heidelberg
London Paris Tokyo Hong Kong

Hans Sünkel
Graz University of Technology
Institute of Mathematical Geodesy
Rechbauerstrasse 12
A-8010 Graz
Austria

Trevor Baker
Bidston Observatory
Birkenhead, Merseyside L43 7RA
United Kingdom

Series Editor
Ivan I. Mueller
Department of Geodetic Science & Surveying
The Ohio State University
Columbus, OH 43210-1247
USA

For information regarding previous symposia volumes contact:
Secretáire Général
Bureau Central de l'Association Internationale de Géodésie
138, rue de Grenelle
75700 Paris
France

Library of Congress Cataloging-in-Publication Data
Sea surface topography and the geoid / Hans Sünkel, Trevor Baker, editors.
p. cm.—(International Association of Geodesy symposia ; symposium 104)
Proceedings of a symposium held within the General Meeting of the IAG in Edinburgh, Scotland, August 10–11, 1989.
Includes bibliographical references.
1. Earth—Figure—Congresses. 2. Sea level—Congresses.
I. Sünkel, Hans. II. Baker, Trevor. III. International Association of Geodesy. General Meeting (1989 : Edinburgh, Scotland)
IV. Series.
QB283.S45 1990
526'.1—dc20 90-9523

Printed on acid-free paper.

ISBN-13: 978-0-387-97268-8

Camera-ready copy provided by the editors.

9 8 7 6 5 4 3 2 1

ISBN-13: 978-0-387-97268-8 e-ISBN-13: 978-1-4684-7098-7
DOI: 10.1007/978-1-4684-7098-7

Foreword

A General Meeting of the IAG was held in Edinburgh, Scotland, to commemorate its 125th Anniversary. The Edinburgh meeting, which attracted 360 scientific delegates and 80 accompanying persons from 44 countries, was hosted jointly by the Royal Society, the Royal Society of Edinburgh and the University of Edinburgh.

The scientific part of the program, which was held in the Appleton Tower of the University, included the following five symposia:

Symposium 101	Global and Regional Geodynamics
Symposium 102	GPS and Other Radio Tracking Systems
Symposium 103	Gravity, Gradiometry and Gravimetry
Symposium 104	Sea Surface Topography, the Geoid and Vertical Datums
Symposium 105	Earth Rotation and Coordinate Reference Frames

All together there were 90 oral and 160 poster presentations. The program was arranged to prevent any overlapping of oral presentations, and thus enabled delegates to participate in all the sessions.

The 125th Anniversary Ceremony took place on August 7, 1989, in the noble surroundings of the McEwan Hall where, 53 years earlier, Vening-Meinesz gave one of the two Union Lectures at the 6th General Assembly of the IUGG. The Ceremony commenced with welcome speeches by the British hosts. An interlude of traditional Scottish singing and dancing was followed by the Presidential Address given by Professor Ivan Mueller, on 125 years of international cooperation in geodesy. The Ceremony continued with greetings from representatives of sister societies, and was concluded by the presentation of the Levallois Medal to Professor Arne Bjerhammar. The 125th Anniversary was also commemorated by an exhibition entitled *The Shape of the Earth*, which was mounted in the Royal Museum of Scotland. An abbreviated version of the President's speech and the list of all participants are included in the proceedings of Symposium 102.

A social program enabled delegates to experience some of the hospitality and culture of both Edinburgh and Scotland, as well as provided an opportunity to explore the beautiful City of Edinburgh and the surrounding countryside.

A Scottish Ceilidh on the last night concluded a pleasant week, which was not only scientifically stimulating, but also gave delegates and accompanying persons an opportunity to renew *auld acquaintances* and make new ones.

The International Association of Geodesy and the UK Organizing Committee express their appreciation to the local organizers of the General Meeting, especially to Dr. Roger G. Hipkin and Mr. Wm. H. Rutherford, for their tireless efforts in running the meeting to its successful conclusion.

Commencing with these symposia the proceedings of IAG-organized scientific meetings will be published by Springer Verlag Inc., New York from author-produced camera-ready manuscripts. Although these manuscripts are reviewed and edited by IAG, their contents are the sole responsibility of the authors, and they do not reflect official IAG opinion, policy or approval.

V. Ashkenazi
A. H. Dodson
UK Organizing Committee

Ivan I. Mueller
President, International
Association of Geodesy

Preface

Symposium 104 aimed at the dissemination of recent advances and techniques in precise geoid determination, discussing their implication for geophysical and oceanographic research, and identifying and formulating a goal on which geoid research should focus in the future.

The Symposium was divided into four sessions with a total of 30 oral and poster presentations. The oral presentations were given on August 10; the poster session on August 11, 1989.

In Session 1 on *The Geoid*, the latest Goddard Earth Model GEM-T3 which is complete to degree and order 50 and based on tracking data of 31 satellites, surface gravity data, and satellite radar altimetry data was presented. The enormous progress in GPS applications has demonstrated its strong impact on geoid determination both by a combination of GPS and INS enabling an almost continuous mapping of a geoid profile and as a very reliable means to control the geoid by a combination of GPS heights with terrestrial heights. A great number of papers dealt with local geoid determinations in various parts on earth. Least squares collocation, supplemented by a remove-restore technique and by sophisticated and frequency domain techniques, is generally recognized as the most powerful tool and has become widely applied for the determination of the gravity field in general and of the geoid in particular.

In Session 2 on *Sea Surface Topography* a large-scale study of an integrated determination of the dynamic sea surface topography, satellite orbit, and the tidal field, based on GEOS-3, SEASAT, and GEOSAT data was presented, providing a precision of 10 cm over basinwide scales and showing an excellent agreement with climatologically induced phenomena. Much attention is being paid to the investigation of the combination of gravity anomalies on land and altimetry derived gravity disturbances over sea in the context of the altimetry-gravimetry problem, and on the simultaneous estimation of the gravity field and the sea surface topography from satellite altimeter data.

In Session 3, *The Vertical Datum* was treated by several authors, providing an overview of three approaches for the connection of vertical datums between continents, based on (1) global gravity field models, local gravity surveys, and levelings, (2) on the solution of a geodetic boundary value problem, and (3) based upon satellite altimetry. Optimal advantage can be taken of the very accurate geometric heights derived from space techniques such as SLR, VLBI, GPS, DORIS, and PRARE, combined with accurate geoid solutions and leveling data, tidal records and sea surface models derived from satellite radar altimetry, for a consistent estimation of an accurate sea surface topography model, the monitoring of mean sea level change, the calibration of radar altimeters, etc. It was demonstrated that global tide gauge sea level data bears very valuable information on the global 18.6 year lunar tide and the 14-month pole tide with amplitudes and phases consistent with a global equilibrium response. Several papers dealt with a comparison between geodetic and oceanographic leveling results and differences which are now being resolved using GPS data, based on precise ephemerides, and a very accurate geoid solution.

In the joint-view Session 4, a simultaneous determination of a general ocean circulation model and an earth gravity field model in terms of spherical harmonics complete up to degree and order 10 for the sea surface topography and complete to degree and order 50 for the geopotential model, derived from SEASAT and GEOSAT altimeter data, Doppler data and SLR data, was presented. Much attention was paid to the highlights of the joint NASA/CNES TOPEX-POSEIDON mission to be flown in late 1991.

Most of the oral and/or poster contributions have been submitted for publication in these proceedings. The convenors would like to express their sincere thanks to all colleagues who contributed to the success of this Symposium.

H. Sünkel T. Baker

Contents and Program

The Vertical Datum

Miscellaneous

GEOID DETERMINATION OVER BASIN-WIDE SCALES USING A COMBINATION OF SATELLITE TRACKING, SURFACE GRAVITY AND ALTIMETER OBSERVATIONS

J. G. Marsh, F. J. Lerch
NASA/Goddard Space Flight Center

S. M. Klosko, T. L. Engelis, G. B. Patel,
J. W. Robbins, R. G. Williamson
ST Systems Inc.

ABSTRACT

A spherical harmonic model of the Earth's gravitational field has been developed using tracking data of 31 satellites, surface gravimetry, and satellite radar altimetry. The model is complete to degree and order 50 and provides a simultaneous recovery for invariant and tidally induced gravitational effects. The ocean tidal model consists of 600 background tidal terms with 90 coefficients being adjusted. The altimeter data are utilized as tracking observations of the ocean surface and provide for a simultaneous recovery of a model describing the stationary sea surface topography complete to degree and order 10 in spherical harmonics. Comparisons between satellite-only models and those obtained with the addition of altimetry/gravimetry find an improvement in geoid representation of more than a factor of two, extending even to the model's longest wavelengths. The stability of the geoid recovery has been assessed using subset solutions. For wavelengths of 2000 km or longer, the combination model provides geoid definition of 6 cm rms accuracy.

INTRODUCTION

One of the major goals of satellite oceanography is to determine the ocean circulation patterns on a global basis. This includes (1) the mean circulation over the broad ocean basins; (2) fluctuations about the mean which reveal phenomenon like seasonal mass transport and climatological changes like the "el Nino"; and (3) long term secular changes in the sea surface which can be used to detect worldwide changes in climate such as the "greenhouse" effect. Altimeter satellites are extremely valuable for supporting the acquisition of global data for ocean monitoring which are well resolved temporally. For investigating cases (1) and (3), the surface mapping provided by altimetry must be supplemented by accurate knowledge of the earth's gravitational field in order to isolate the oceanographic signals present in the sea surface.

Improved gravitational models are important for oceanographic investigations using spaceborne altimeter systems. Accurate geopotential models (both static and temporally resolved) are required to compute the satellite ephemeris. Knowledge of the satellite's height is indispensable for determining the absolute location of the ocean surface with respect to the geocenter. The sea surface height is dominated by the geoid; again accurate gravitational and tidal models are required to determine the departure of the ocean surface attributable to geostrophic flow from that of the equipotential surface. The direct inclusion of altimeter data in the definition of the gravitational field provides a new and important resource for the improvement of orbit and gravitational modeling effects. However, the utilization of these observations in geopotential solutions is complicated by the need to successfully accommodate the non-geoidal and incompatible bandwidth signals contained in the altimetry.

Significant progress has been made in developing models of the terrestrial gravity field using satellite data. With the advent of improved satellite tracking technologies like satellite laser ranging, space-based laser targets like the Lageos and Starlette satellites, and the deployment of a global network of tracking systems, these "satellite-only" models are now capable of defining the long wavelength geoid to near-decimeter levels. Combination models have been developed which augment tracking data with surface information which improves the modeling of the short wavelength geoidal features which are not well sensed by the attenuated gravitational effects at satellite altitude. These models have employed global gravimetry which has been reduced to form mean gravity anomalies blocks; these mean anomalies have been used to compute normal equations for the recovery of spherical harmonic gravity fields (Pavlis, 1988). Satellite altimetry has also been used directly in these combination models to better resolve the geoid and the radial evolution of satellite orbits as they overfly the ocean surface. This paper reviews the progress being made at NASA/Goddard Space Flight Center (GSFC) in the modeling of the terrestrial gravity field from both satellite and surface information.

GEM-T2: A MODEL DEVELOPED FROM TRACKING OBSERVATIONS

Goddard Earth Model (GEM)-T2 (Marsh et al., 1989b) has been computed using satellite tracking observations acquired on 31 different satellite orbits. This solution is compared to its predecessor, GEM-T1 (Marsh et al., 1988) in Table 1. GEM-T2 contains terms in the spherical harmonic expansion which are complete to degree and order 36, with selected orders containing adjusted values out to degree 50. This model now nearly exhausts the historical satellite data set available for gravitational field development; it includes laser, optical, Doppler, satellite-to-satellite range-rate, and S-Band average range-rate observations. While additional observations are available on the satellites represented within the model, all existing important satellites for geodetic evaluation are now incorporated. The accuracy of this model obtained from its calibrated error covariance is shown in Figure 1 and is compared with GEM-T1. Figure 1 displays the accuracy assessment in terms of the RMS coefficient uncertainty by degree. The calibration procedure used to obtain realistic error estimates for the GEM models is fully described in Lerch et al., (1988). The development of an automated procedure for determining optimal data weights simultaneously with a properly calibrated field (which has been used for the first time in the determination of GEM-T2) is presented in Lerch (1989).

Orbital accuracies are a central concern and one which motivates the development of improved gravitational models. One of the best ways for assessing the performance of these models is testing them against a broad spectrum of satellite orbits. Table 2. compares the performance of recent GEM models based on the RMS of fit they yield when computing orbital ephemerides. The progress is quite notable. However it is increasingly more difficult to improve these RMS values as gravity field modeling is diminished as the overwhelming contributor to the misclosure of model and observations in these orbital computations. Many additional force model, environmental, and tracking station motion effects require improved modeling to accomplish this objective.

PGS-3520: A COMBINATION MODEL WITH GEM-T2, SURFACE GRAVIMETRY AND SEASAT ALTIMETRY

The gravitational information sensed by satellite tracking systems is complemented by that obtained from surface gravimetry and satellite altimetry. While the earlier has a reduced sensitivity to short wavelength geoid features due to attenuation, the surface data and altimetry contain a rich spectrum of short wavelength geopotential information. In addition, when altimetry is used to monitor the radial component of the orbit trajectory, it provides information which is complementary to the largely along track sensitivity found with traditional ground-based tracking systems. Figure 4 compares the gravity model uncertainties obtained from the "satellite-only" GEM-T2 model, and a model computed using only altimetry and surface gravimetry.

Table 1.

Improvements in GSFC "Satellite Only" Gravitational Models

Data	GEM-T1 (1987)	GEM-T2 (1989)
No. of Observations	792k	2386k
No. of Arcs	581	1132
No. of Satellites	17	31
Amount of LAGEOS data	4.8 years	7.2 years
Size of static gravitational model	36 x 36 complete	36 x 36 complete with 600 additional terms to (50,43)
Size of adjusting tide model	66 coeffs.	90 coeffs.
RMS geoid error		
• at 36 x 36	155 cm	105 cm
• complete model	--	141 cm

Figure 1.

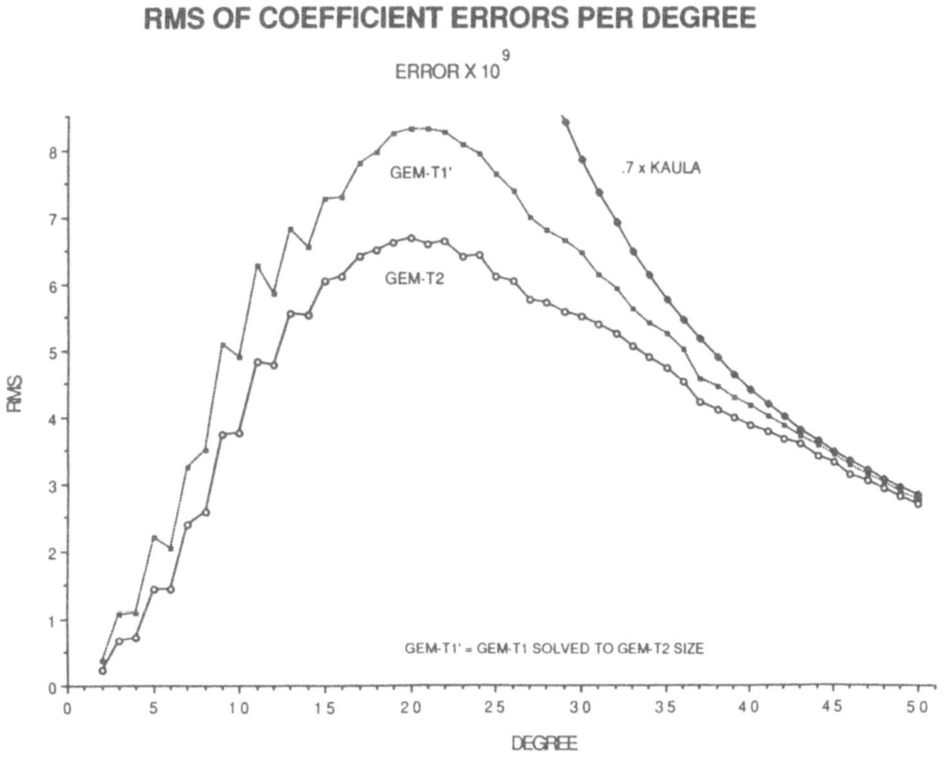

When altimetry is utilized directly in geopotential solutions, provision must be made for the long wavelength departure of the ocean surface from the geoid due to the quasi-stationary nature of the sea surface topography over basin-wide scales. PGS-3520 is a combination model based on GEM-T2, global surface gravimetry and SEASAT altimetry. It is complete to degree and order 50. In forming a combination solution containing altimetry, we simultaneously solved for a spherical harmonic model for the stationary dynamic height field complete to degree and order 10. We also extensively masked, corrected and optimally subsampled the altimetry as described in Marsh et al., (1989a) for the PGS-3337 solution, before these data were included in the PGS-3520 field. The result of combining these data are shown in Figure 4 where the RMS coefficient uncertainty by degree is also shown for PGS-3520.

The improvement in geoid accuracy from recent gravitational models is quite significant and is especially important for applications requiring accurate geoid modeling at long and intermediate wavelengths over the oceans. Figures 2a and 2b present the estimated geoid uncertainty obtained from a covariance propagation of model errors. These figures show the estimate of the commission error in the GEM-T2 and PGS-3520 solutions complete to degree and order 36 (which is the level at which GEM-T2 is complete). This truncation represents approximately 500 km surface wavelengths. The uncertainties shown for GEM-T2 (while significantly improved over the 160 cm RMS value obtained for GEM-T1), are zonally banded indicating that there is not a high level of geographic discrimination obtained in gravitational solutions dependent on satellite tracking observations. The orbit perturbations which are predominantly sensed by tracking data are the m-daily and resonance perturbations, and these are globally sampled gravitational effects arising from the earth's rotation and the satellite's mean motion. This situation changes both in modeling accuracy and also in character as surface/altimeter data are added to the solution. In Figure 2b, the geoid uncertainty over the ocean areas which are mapped by the altimeter is found to be minimized. And it is now the regions where the surface representation is poorest (e.g. Asia, the USSR, the poles) that one finds the largest geoid uncertainties. The characteristic of the error is also changed in altimeter mapped regions, where the uncertainty is seen to have little structure.

Figures 3a and 3b compare the geoid uncertainty for the longest wavelength 10x10 portion of these fields. Again, GEM-T2 is a significant improvement over earlier satellite only gravity models. At the indicated 10-20 cm level of geoid uncertainty, there is now a favorable basis for resolving the dynamic height field over similar wavelengths which has a signal strength of approximately 1m. The combination model, PGS-3520, shows significant improvement over GEM-T2, even at these long wavelengths. This is a product of the strength of the altimetry for improving the orbits; there is dramatic decorrelation in the model when altimetry is used which penetrates down to the longest wavelength terms; and these data directly contribute by mapping the shape of the surface across the ocean basins. The simultaneous adjustment of a 10 x 10 degree model for dynamic height in PGS-3520 has not significantly dampened the strength and importance which altimetry adds to the solution. By using the strength of the geoid determination found in our combination models, the dynamic height recovery has been extended to degree 20 in test GSFC solutions.

Earth modeling along the lines of PGS-3520 will be pursued with the addition of altimeter data from GEOSAT and GEOS-3 to produce GEM-T3. This work is expected to be completed in early 1990.

MODEL CALIBRATION

The GEM models are now subjected to a comprehensive calibration procedure. This now automated approach yields a well calibrated solution through the optimal weighting of the observation subsets. This calibration and data weighting system is based on a method which tests the model against solutions lacking specific data subsets. Through the comparison of the differences in the coefficients (ΔC) obtained between the subset and complete solution

GEOID HEIGHT STANDARD DEVIATIONS USING GEM-T2 COVARIANCES TO 36 X 36

CONTOUR INTERVAL 5 CM. RMS = 105 CM.

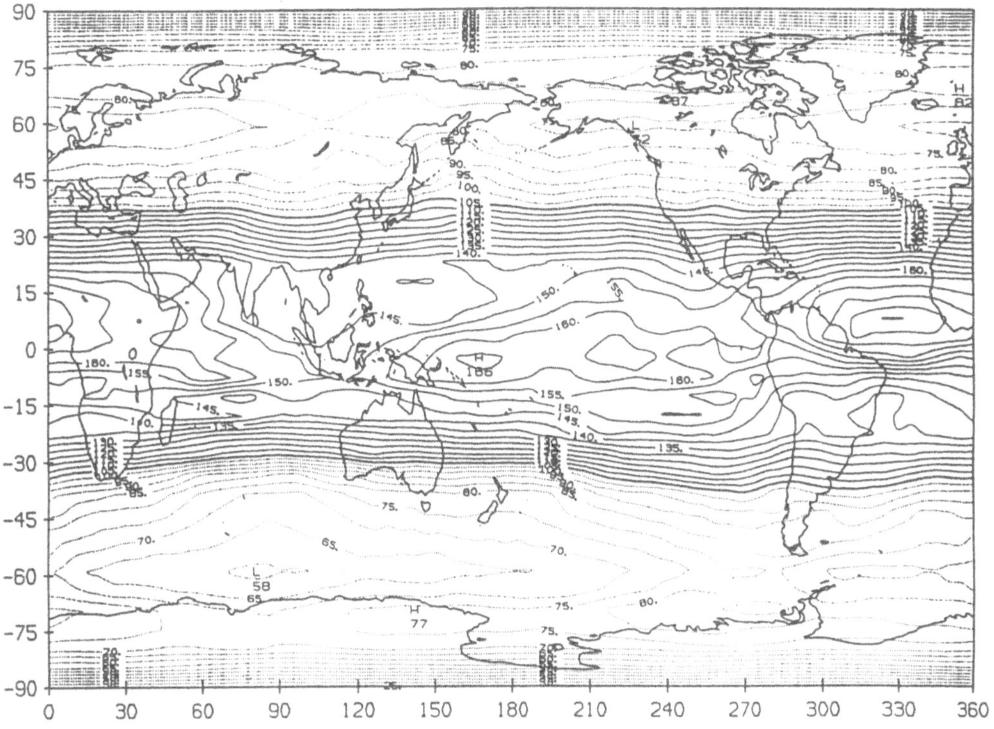

Figure 2a.

GEOID HEIGHT STANDARD DEVIATIONS USING PGS3520* COVARIANCES TO 36 X 36

CONTOUR INTERVAL 5 CM. RMS = 35 CM.

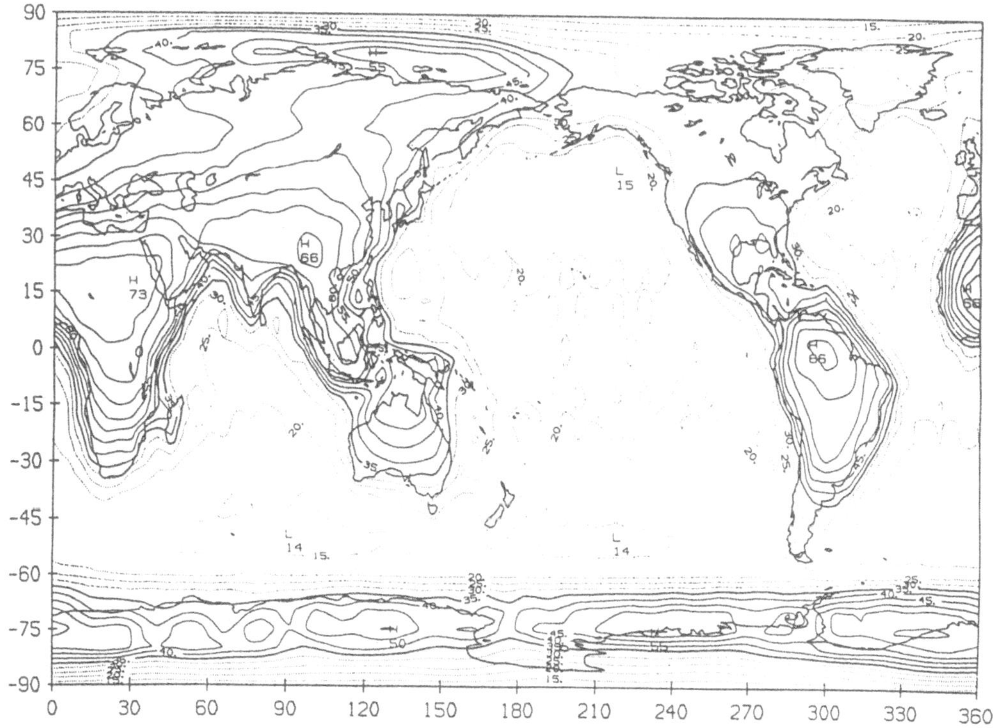

* PGS3520 = GEM-T2 + SURFACE GRAVITY + SEASAT ALTIMETER

Figure 2b.

GEOID HEIGHT STANDARD DEVIATIONS USING GEM-T2 COVARIANCES TO 10 X 10

CONTOUR INTERVAL 2 CM. RMS = 15 CM.

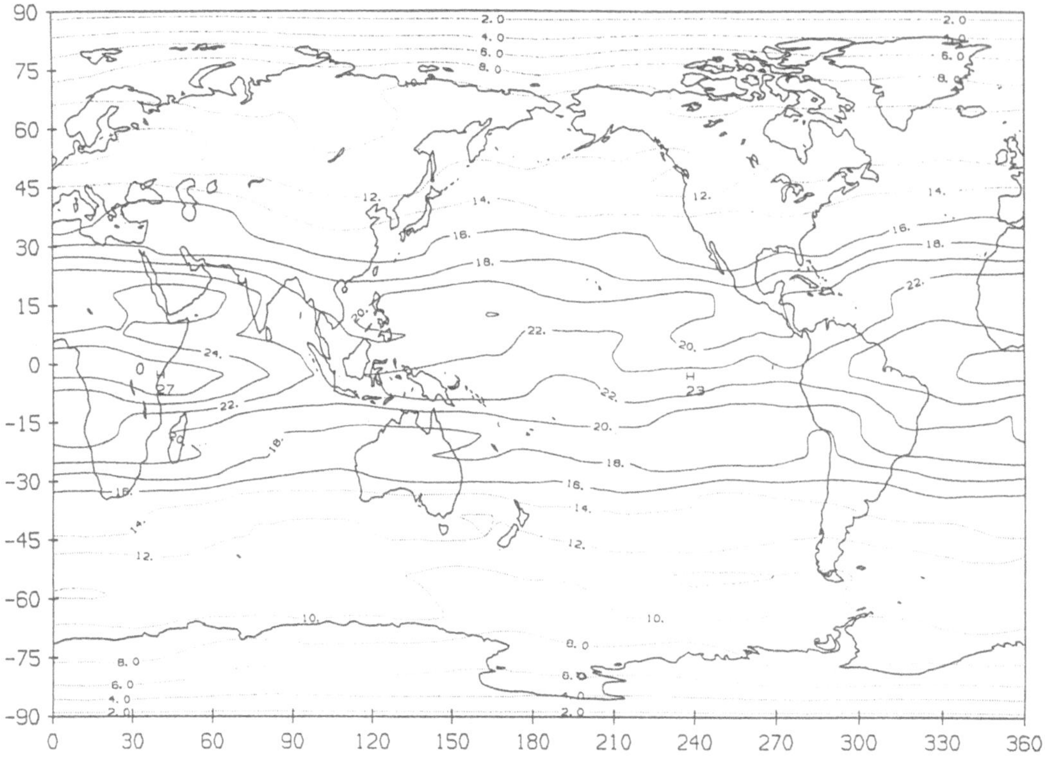

Figure 3a.

GEOID HEIGHT STANDARD DEVIATIONS USING PGS3520* COVARIANCES TO 10 X 10

CONTOUR INTERVAL 1 CM. RMS = 6 CM.

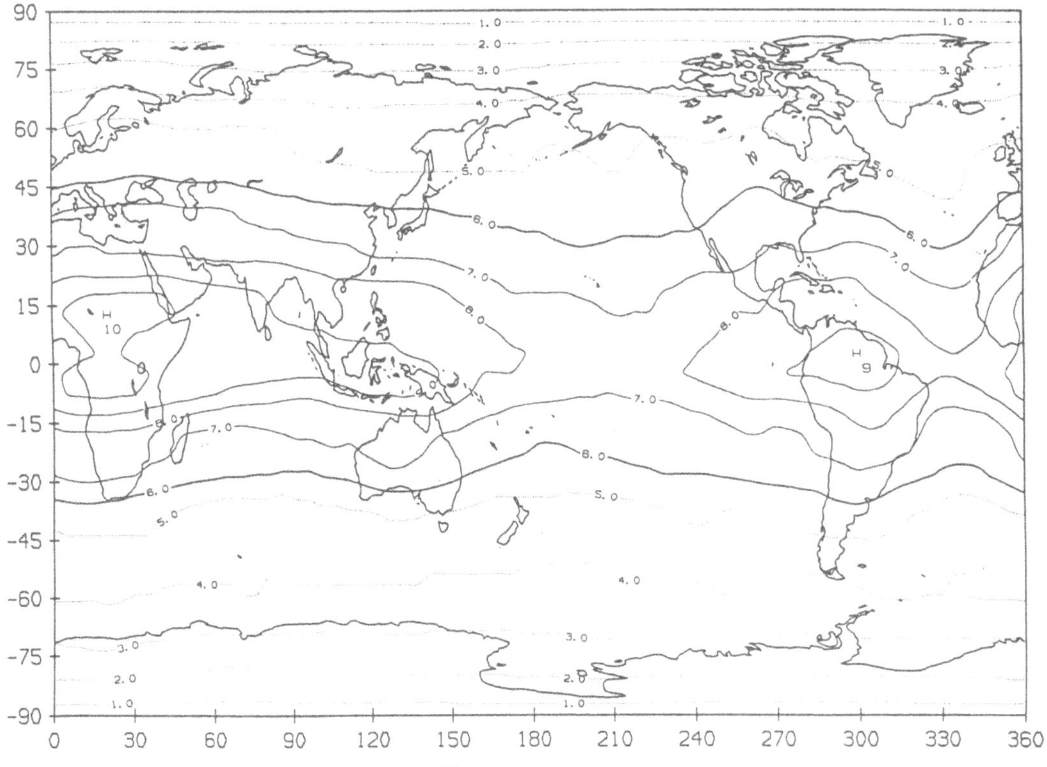

Figure 3b.

* PGS3520 = GEM-T2 + SURFACE GRAVITY + SEASAT ALTIMETER

Table 2.

Orbit Accuracy Assessments of Satellite Fields Using Test Arcs

(rms of fit)

Gravity Field	LAGEOS (m)	AJISAI (m)	STRLT (m)	BE-C (m)	GEOS1 (m)	GEOS2 (m)	GEOS3 (m)	NOVA (cm/s)	ALTIM X-OVER (m)
GEM-9 (1979)	.333	.951	1.16	.873	1.26	1.18	1.72	0.95	3.65
GEM-L2 (1981)	.199	.797	1.00	.893	1.07	1.09	1.87	0.79	6.42
GEM-T1 (1987)	.069	.181	.172	.396	.387	.655	.693	0.44	0.65
GEM-T2 (1989)	.066	.151*	.102	.334	.316	.667	.249	0.37*	0.57
NOISE FLOOR	.038	.038	.040	.102	.206	.343	.101	.335	----

* Satellite now in model

Figure 4.

RMS OF COEFFICIENT ERROR PER DEGREE

SATELLITE TRACKING DATA SOLUTION
VS.
ALTIMETER + SURFACE GRAVITY SOLUTION

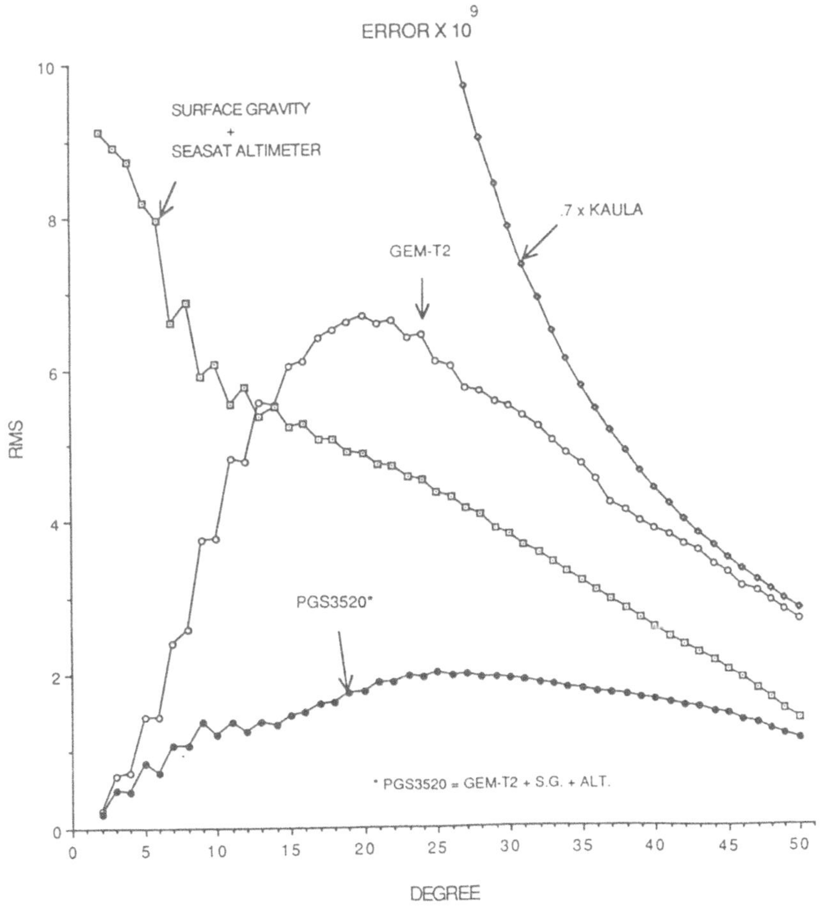

Table 3.

Summary of Calibration Factors k for GEM-T2 Using Subset Solutions and Other Measures of Subset Field Performance

Data Subset Omitted from GEM-T2	k (Overall Calib.) Factor	Estimated Geoid Ht. Error (m)	Comparison with Altim Grav. Anom. (mgals**2)
none	----	140.5	12.5
Ajisai	0.79	141.3	12.6
GEOSAT	0.81	145.8	12.9
NOVA	0.90	146.2	12.9
Peole, D1D, D1C	0.87	159.0	12.8
Lageos	0.87	141.4	12.6
pre-1980 GEOS 1,2,3, BEC	1.01	146.2	13.6
Oscar	1.13	141.1	12.6
Starlette	0.96	150.2	14.1
SEASAT	0.94	141.1	12.7
Landsat	0.92	142.8	12.7
1980 GEOS-1	1.05	141.3	12.5
1980 GEOS-3	0.96	143.2	12.6
Optical Data	0.92	147.1	13.5
GEOS-3/ATS-6 SST	0.66	140.7	12.5

compared with an error projection of the expected size of the coefficient differences $\sigma(\Delta C)$ obtained from the respective solution covariances, a calibration factor is determined by:

$$|\Delta C| = k\ \sigma(\Delta C)$$

where ideally the calibration factor, k = 1. All of the major data subsets are tested and the solution data weights are iterated until a calibration factor near unity is obtained for each. Table 3 reviews the calibration factors obtained in GEM-T2. In the unique case of the GEOS-3/ATS-6 data set, the calibration factor (data weight) was deliberately permitted to deviate from unity reflecting our desire to downweight these data. This decision was based upon test against independently derived gravity anomaly blocks developed from altimeter data (Rapp, 1985). The comparison of the subset models with these surface anomalies is also shown in Table 3. as the residual (in $mgal^2$) between the model and the surface data. Note, in all cases, the GEM-T2 model outperforms the subset models indicating that each data set is contributing to the overall accuracy of the field.

SUMMARY

Improvements in gravity modeling have been derived through the combination of satellite tracking, altimetry and surface gravimetry observations. The long wavelength geoid now has sub-decimeter accuracy levels and supports the long wavelength separation of geoidal and oceanographic signals sensed by altimeter systems. These models also yield improved capabilities for determining the satellite ephemerides which has wide ranging implications for all aspects of satellite geodesy. The models are still limited by missing local data over large, mostly continental, geographical areas. Areas lacking surface coverage in the solution now are most uncertain for geoid modeling at all wavelengths. The historical satellite tracking data set still lacks optimal distribution of satellite orbital characteristics; especially lacking are laser tracked satellites at low inclination. Also improvements in many areas of force and measurement modeling are required to fully exploit precise laser tracking for geopotential modeling improvements. Without a dedicated geopotential mission, global model accuracies are not likely to improve by more than an additional factor of 3 or 4 beyond present knowledge of the fields at long and intermediate wavelengths. It is hoped that new technologies, like the nearly continuous tracking possible for lower satellites using GPS, or the space-based gradiometer flown on Aristoteles, will significantly improve this prospect.

REFERENCES

Lerch, F.J., J.G. Marsh, S.M. Klosko, E.C. Pavlis, G.B. Patel, D.S. Chinn, C.A. Wagner, An Improved Error Assessment for the GEM-T1 Gravitational Model, NASA TM-100713, November 1988.

Lerch, F. J., Optimum Data Weighting and Error Calibration for Estimation of Gravitational Parameters, NASA TM-100737, July, 1989.

Marsh, J.G. et al., A New Gravitational Model for the Earth from Satellite Tracking Data: GEM-T1, J. Geophys. Res., 93, B6, 6169-6215, 1988.

Marsh, J.G., F.J. Lerch, C.J. Koblinsky, S.M. Klosko, J.W. Robbins, R.G. Williamson and G.B. Patel, Dynamic Sea Surface Topography, Gravity and Improved Orbit Accuracies from the Direct Evaluation of SEASAT Altimeter Data, NASA TM-100735, May, 1989a.

Marsh, J.G. et al., The GEM-T2 Gravitational Model, NASA TM-100746, September, 1989b.

Pavlis, N.K., Modeling and Estimation of Low Degree Geopotential Model from Terrestrial

Gravity Data, The Ohio State University Department of Geodetic Science and Surveying, Report No. 386, Columbus, 1988.

Rapp, R., Detailed Gravity Anomalies and Sea SUrface Heights Derived from GEOS-3/SEASAT Altimeter Data, The Ohio State University Department of Geodetic Science and Surveying, Report No. 365, Columbus, 1985.

GEOID DETERMINATION USING GPS-AIDED INERTIAL SYSTEMS

Elfriede Knickmeyer
Klaus Peter Schwarz
Department of Surveying Engineering
The University of Calgary
2500 University Drive N.W.
Calgary, Alta., CANADA T2L 2J5

ABSTRACT

Inertial systems have been used for about fifteen years to determine the anomalous gravity vector at discrete points using zero velocity update measurements at vehicle stops. This paper discusses a kinematic method of determining continuous geoid profiles from a combination of INS and GPS measurements. The major error sources, namely gyro drift, accelerometer bias variation, and GPS noise are modelled separately and their effect on the resulting relative geoid is studied.

1. INTRODUCTION

In a previous paper, a kinematic method of determining continuous geoid profiles from a combination of INS and GPS measurements has been proposed, see Schwarz (1987). Since INS measurements are sensitive to the local gravity field, while GPS measurements are not, position or velocity differences between the two system outputs can be used to derive the gravity disturbance vector along the trajectory. Relative geoid profiles between two points can then be obtained by integrating the deflections of the vertical along the vehicle path.

A first analysis of this idea showed that the accuracy of the resulting geoid is dependent on three parameters: the vehicle velocity, the shortest wavelength of the gravity field to be recovered, and the accuracy with which the difference between INS and GPS position or velocity can be obtained. The analysis was done in terms of wavelengths of the gravity field and showed that with current measurement accuracies, it would be very difficult to recover wavelengths smaller than 10 km. Since it is known that these wavelengths contribute very little to the local geoid, the method seemed feasible and was further pursued.

This paper extends the previous analysis in three points. First, instead of looking at the accuracy of wavelength recovery, it estimates the total accuracy of the resulting local geoid. This is done by assuming that a good geopotential model is available as a first approximation, and that the remaining energy of the gravity field can be adequately modelled by a Gauss-Markov process, for a discussion see Vassiliou and Schwarz (1987). Second, the analysis is refined by separately modelling the major error sources, namely gyro drift, accelerometer bias variation, and GPS noise. Third, the relative geoidal height is directly modelled as a component of the state vector, i.e. the integration of the deflections of the vertical is performed implicitely. In this way, the accuracy of the resulting relative geoid can be directly related to the different error sources.

2. PRINCIPLE OF RELATIVE GEOID DETERMINATION BY GPS-INS

To discuss the salient features of the method, the single axis inertial navigator will be used as an example. This fictitious device allows a simple treatment of the problem and gives results which show essentially the same characteristics as the horizontal channels of a three axes navigator over time spans of one to two hours. The single axis system consists of a gyro and an accelerometer mounted on a level platform. The gyro defines the reference frame for the acceleration measurements. The accelerometer measures the combined effect of kinematic and gravitational acceleration, the so-called specific force

$$f = \ddot{r} - \xi g \tag{1}$$

where r is the position, ξ is the deflection of the vertical and g is the magnitude of the gravity vector. The two accelerations ξg and $\ddot{r}$ cannot be distinguished by measuring f alone at discrete points. To separate them, an independent measurement of either ξg or $\ddot{r}$ must be available. For a moving vehicle, the first possibility can be implemented by using a gravity gradiometer, the second by using GPS. In the following, we will only consider the case where GPS is used as an aiding system.

Rewriting equation (1), the deflection can be obtained from

$$\xi(t) = \frac{\ddot{r}(t) - f(t)}{g(t)} \quad . \tag{2}$$

A first examination of this equation shows that it will be very difficult to match the sensor accuracy of a well-calibrated high performance accelerometer, which measures f with an accuracy of a few mGals, by acceleration measurements $\ddot{r}$ derived from GPS. Differentiating GPS velocity measurements, which in the best of cases may be good to about 5 mm/s, will result in acceleration errors of several hundred mGals. Even when eliminating the high frequency part of the noise spectrum by filtering, it will be very difficult to reach the accuracy required for airborne gravimetry as Kleusberg et al. (1989) have recently shown. This is not so for geoid determination as will be shown in the remainder of this paper.

Using astronomic levelling, the deflections can be integrated to give relative geoid heights

$$\Delta N(r) = - \int_{r(P1)}^{r} \xi(\bar{r})\, d\bar{r} \qquad (3a)$$

or

$$\Delta N(t) = - \int_{t(P1)}^{t} \xi(r(\tau)) \cdot v(\tau)\, d\tau \qquad (3b)$$

where P_1 is the starting point of the traverse. This formula will be used for the description of the relative geoidal height in the state vector. The mapping from time to space is possible because position and velocity are determined as time variables. Since the deflections are measured on the surface of the earth, the integrated quantities are differences in height anomaly. In relative geoid determination, the difference between height anomalies and geoidal undulations is usually very small and will be disregarded in the sequel.

3. STATE SPACE REPRESENTATION

The error behaviour of systems of this kind can be described using the state space model

$$\dot{\mathbf{x}}(t) = \mathbf{F} \cdot \mathbf{x}(t) + \mathbf{G} \cdot \mathbf{w}(t) \qquad (4)$$

where x is the state vector, **F** is the dynamics matrix, **G** is the shaping filter and **w** is unity white noise. The state space model for the single axis inertial navigator is well documented, e.g. in Schwarz (1983). Its linearized form is

$$\frac{d}{dt}\begin{pmatrix} \delta r \\ \delta v \\ \varepsilon \end{pmatrix} = \begin{pmatrix} 0 & 1 & 0 \\ 0 & 0 & -g \\ 0 & 1/R & 0 \end{pmatrix} \bullet \begin{pmatrix} \delta r \\ \delta v \\ \varepsilon \end{pmatrix} \qquad (5)$$

where δr, δv, and ε are the corrections to the unknown position, velocity, and attitude, respectively, g is the magnitude of gravity, and R is the radius of a local sphere approximating the earth.

In order to include the deflections of the vertical, model (5) has to be augmented. Different approaches to include gravity field parameters in the state space model are given in Jordan et al (1981), Schwarz and Wei (1988), Eisfeller and Spietz (1989). For this study it is considered sufficient to use a simple Gauss-Markov type model for the deviations of the deflections of the vertical from a mean value or some other model. Vassiliou and Schwarz (1987) have shown that among these a second-order Gauss-Markov model is best suited for modelling the local geoid. Test calculations showed that there is little difference between a first order and a second order model in this specific case. Therefore the simpler first-order model was used to do the numerical computations.

The state-space formulation of a Gauss-Markov process of first order is (Brown, 1983)

$$\frac{d\delta\xi(r)}{dr} = -\beta' \cdot \delta\xi(r) + \sqrt{2\sigma^2\beta'} \cdot w(r) \tag{6a}$$

where σ^2 is the mean square value, and β' is the inverse of the correlation length.

After a variable transformation from r to t, which is done via the observed velocity, it reads

$$\delta\dot{\xi}(t) = -\beta' \cdot v \cdot \delta\xi(t) + \sqrt{2\sigma^2\beta' v} \cdot w(t) \tag{6b}$$

The linearized differential equation for the geoid undulation difference is

$$\begin{aligned} \delta\Delta\dot{N}(t) &= -\xi(t) \cdot v(t) \\ &= -\xi(t) \cdot v_0(t) - \xi(t) \cdot \delta v(t) \end{aligned} \tag{7}$$

where v_0 is the approximate velocity, and δv is the velocity residual. The second term can be neglected because $\delta v << v_0$.

So far, errorless sensors have been assumed. In reality, the measurements contain sensor errors. For the INS, the main errors are biases of the accelerometers and gyros. For gravity field recovery the biases need to be calibrated immediately before and after the run. Thus, the most important factor is the stability of the biases over the period of the run. The residual gyro and accelerometer biases will also be modelled as first-order Gauss-Markov processes. This is still a simplification for the gyro model because even after calibration it should contain a random walk component. The complete state space model is then of the form

$$\frac{d}{dt}\begin{pmatrix} \delta r \\ \delta v \\ \varepsilon \\ \delta\xi \\ \delta\Delta N \\ b_a \\ b_g \end{pmatrix} = \begin{pmatrix} 0 & 1 & 0 & 0 & 0 & 0 & 0 \\ 0 & 0 & -g & g & 0 & 1 & 0 \\ 0 & 1/R & 0 & 0 & 0 & 0 & 1 \\ 0 & 0 & 0 & -\beta_\xi & 0 & 0 & 0 \\ 0 & 0 & 0 & -v_0 & 0 & 0 & 0 \\ 0 & 0 & 0 & 0 & 0 & -\beta_a & 0 \\ 0 & 0 & 0 & 0 & 0 & 0 & -\beta_g \end{pmatrix} \cdot \begin{pmatrix} \delta r \\ \delta v \\ \varepsilon \\ \delta\xi \\ \delta\Delta N \\ b_a \\ b_g \end{pmatrix} + \begin{pmatrix} 0 \\ 0 \\ 0 \\ \sqrt{2(\sigma^2\beta)_\xi} \\ 0 \\ \sqrt{2(\sigma^2\beta)_a} \\ \sqrt{2(\sigma^2\beta)_g} \end{pmatrix} \cdot w \tag{8}$$

where $\beta = \beta' \cdot v$ is the inverse of the correlation length.

The GPS observations are incorporated as updates in a Kalman filter. Rather than modelling individual error sources, the accuracy of position and velocity as determined from GPS is characterized by white noise processes, assuming that large biases have been cancelled by differencing between a stationary and a moving receiver.

4. SENSOR REQUIREMENTS

Selecting the second row of the system equation and solving for $\delta\xi$ yields

$$\delta\xi(t) = \frac{\delta\dot{v}(t)}{g} + \varepsilon(t) - \frac{ab(t)}{g} \tag{9a}$$

$$= \frac{\delta\dot{v}(t)}{g} + \frac{\delta r(t)}{R} + \int_0^t gb(\tau)\, d\tau - \frac{ab(t)}{g} \tag{9b}$$

which is a linearized and extended version of (2).

As $\delta\xi(t)$ changes only slowly, averaging is possible. The averaging time depends on the correlation length of ξ and on the speed of the vehicle.It can be seen, that random bias variations, or their integral, respectively, will be interpreted as variations in ξ.

Assuming constant velocity and constant magnitude of gravity, (9b) can be integrated to obtain relative geoid undulations

$$\delta\Delta N(t) = -\frac{v}{g}\cdot\delta v(t) - \frac{v}{R}\cdot\int_0^t \delta r(\tau)d\tau - v\cdot\int_0^t\int_0^t gb(\tau)\, d\tau\, d\tau + \frac{v}{g}\cdot\int_0^t ab(\tau)d\tau \tag{10}$$

A first look at the equation shows already that the influence of the biases will grow with speed for the first term and with distance travelled for the last three terms. The influence of the gyro bias is, in addition, inversely proportional to the speed. This suggests the existence of an optimal velocity for a certain hardware configuration and gravity field.

In order to compute the effect of biases, the error propagation formulas of the Kalman filter are applied (Gelb, 1974). The filtered variance-covariance matrix is obtained from

$$\mathbf{Q}_{x,f}(t) = \Phi(0,t)\cdot\mathbf{Q}_x(0)\cdot\Phi^T(0,t) + \int_0^t \Phi(0,\tau)\cdot\mathbf{G}\cdot\mathbf{G}^T\cdot\Phi^T(0,\tau)\, d\tau \tag{11}$$

together with the update equations

$$\mathbf{Q}_x^+ = (\mathbf{I}-\mathbf{KH})\cdot\mathbf{Q}_x^- \tag{12a}$$

$$\mathbf{K} = \mathbf{Q}_x^-\mathbf{H}^T(\mathbf{H}\mathbf{Q}_x^-\mathbf{H}^T + \mathbf{Q}_e)^{-1} \tag{12b}$$

where $\mathcal{Q}_x$ and $\mathcal{Q}_e$ are the covariance matrices of state vector and update observations, respectively, **H** is the design matrix, and **K** is the Kalman gain matrix; "-" indicates predicted, "+" updated quantities.

The first part of (11) constitutes the effect due to errors in the initial conditions, the second part is the error due to system noise. The smoothed variance-covariance matrix is finally obtained from

$$\mathcal{Q}_{x,s}(t) = (\mathcal{Q}_f(t)^{-1} + \mathcal{Q}_b(T-t)^{-1})^{-1} \tag{13}$$

where T is the duration of the survey, and f and b indicate forward and backward filtering, respectively.

The effect of the accelerometer and gyro bias can be derived analytically if all other states are neglected (Huddle, 1978). For a one hour survey time this yields the following maximum errors for the deflection and the geoid undulation.

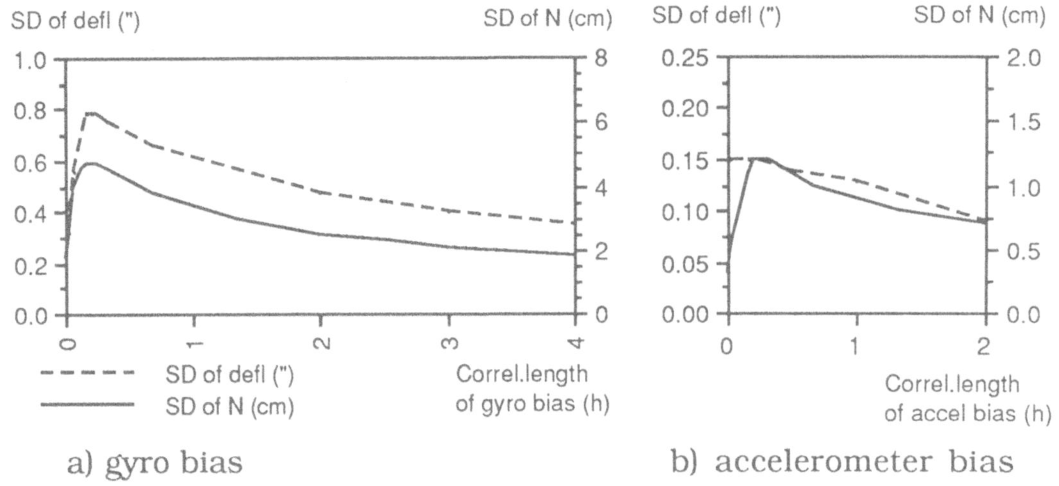

Fig. 1: Maximum geoid and deflection error due to biases

The obtained spectral densities in the deflections of the vertical and the geoid undulations are proportional to the spectral densities of the sensor errors.

In order to investigate the effect of position and velocity updates, a covariance analysis was performed. The main concern in this study is the error due to system noise, not the one due to uncertainty in the initial conditions. In a Kalman filter setting, however, zero initial variances cannot be used, because the filter would not respond any more to update observations. Therefore, small non-zero initial variances are chosen which do not necessarily reflect the actual alignment accuracy. A zero velocity update (ZUPT) is supposed to take place at the beginning and at the end of the survey. In between the vehicle travels for one hour, aided only by GPS.

Table 1: Parameters of covariance analysis

Initial Conditions:				
σ_s:	0.01 m	Deflection model:		Speed: 50 km/h
σ_v:	0.0003m/s	σ_ξ: 5"	ζ': 20km	
σ_ε:	0.2"	Accelerometer model:		Update rate: 0.5Hz
σ_ξ:	0.2"	σ_a: 1 mgal	ζ': 0.5h	
σ_{b_a}:	0.2 mgal	Gyro model:		
σ_{b_g}:	0.0002°/h	σ_g: 0.001°/h	ζ': 2h	Distance: 50 km
σ_N:	0.01m			

The results shown are the maximum errors after smoothing, which occur at the midpoint of the traverse. The effect of position and velocity updates can be seen from Fig.2. The noise in the update measurements has considerably less effect on geoid undulations than on deflections because it is averaged out in the integration process. Whereas the determination of deflections at the sub-arcsecond-level requires extremely high accuracy of the update observations, even modest accuracies will give a relative geoid at the 5 cm level.

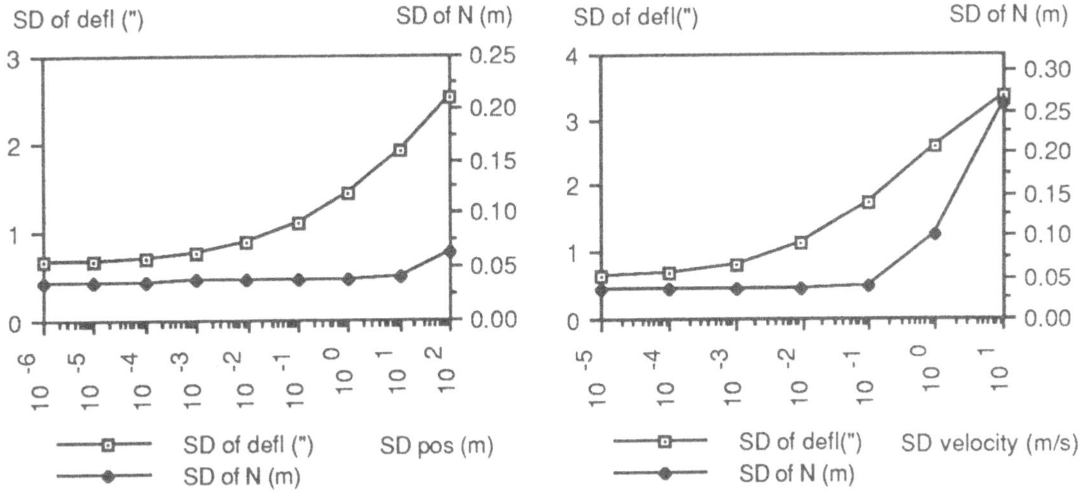

a) position update accuracy b) velocity update accuracy

Fig. 2: Geoid and deflection accuracy as a function of update accuracy

The effect of the update rate can be seen from Fig. 3. The output rates of current receivers (0.2 - 1 Hz) are fully sufficient for the determination of geoid undulations. This holds for both position and velocity updates. Higher update rates would, however, improve deflection determination. Fig. 4 shows the effect on wavelength recovery in the gravity field. Although wavelengths below 10 km cannot be recovered well with this method, the effect on geoid determination is negligible. It can be seen that the results of this study are basically not affected by the chosen correlation length of the deflections, as long as it is larger than 10 to 20 km.

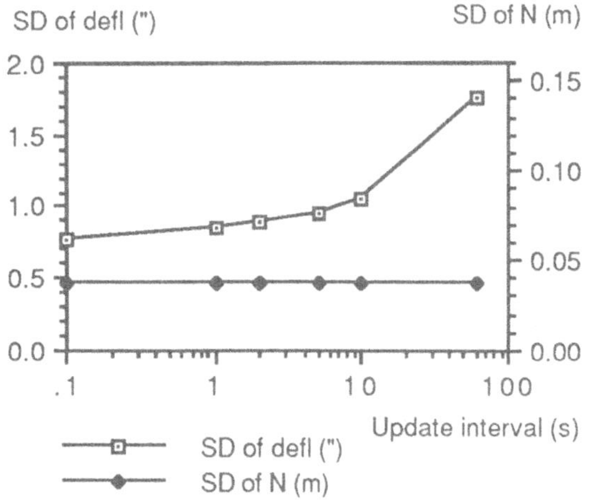

Fig.3: Effect of update rate

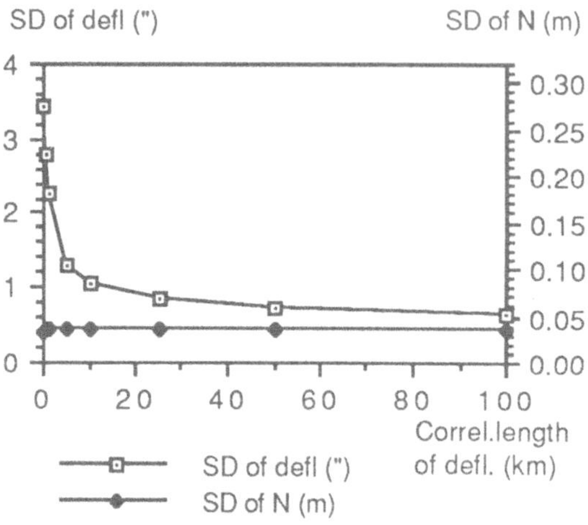

Fig.4: Wavelength resolution

The effect of variations in the gravity field can be seen from Fig. 5 where the horizontal scale gives the root mean square of the deflections. Most of the effect is averaged out in the determination of the geoid. This is not the case for deflection determination.

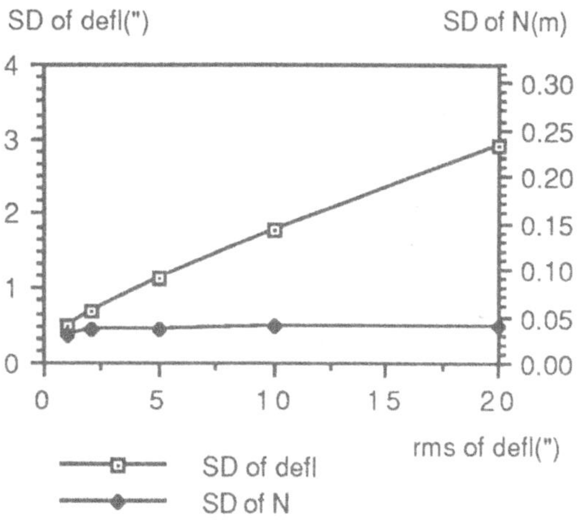

Fig. 5: Effect of variation of the gravity field

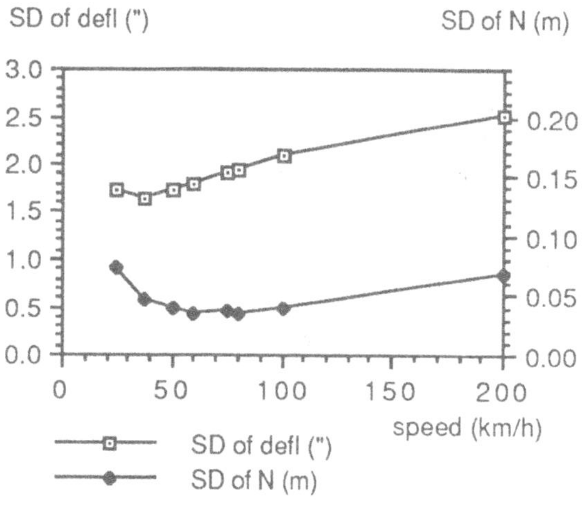

Fig. 6: Effect of vehicle speed

As already mentioned in the context of equation (10), the results are affected by the vehicle speed. Fig. 6 shows that the optimal speed is determined by a tradeoff between the GPS errors, the influence of which grows with velocity, and the INS errors, the influence of which grows with time and therefore decreases when the same distance is covered with a higher speed. Velocities between 40 km/h and 150 km/h are expected to give good results for the geoid. This allows for land vehicle as well as for airborne applications.

5. APPLICATIONS

The method discussed in this paper has a number of important practical applications. The first one, which is especially attractive in conjunction with semi-kinematic GPS methods, is the simultaneous determination of precise three-dimensional positions and orthometric heights without levelling. In this case, the geoidal undulation differences determined by the above method would be subtracted from the ellipsoidal height differences determined by GPS to obtain orthometric height differences. This application would typically be done in a truck and would have major economical advantages in areas with little geodetic information.

The second application is the airborne determination of the local geoid, especially in mountainous areas and in remote regions with little or no gravity coverage. The main advantages of such an approach are speed, economy, more consistent data coverage for surface fitting, and smaller dependence on weather conditions. In this respect, it has all the advantages aeromagnetics has versus geomagnetics on the ground.

A third application that might be feasible is the determination of the difference between geoid and mean sea surface in oceanic regions. This application would obviously be restricted to resolving short wavelength characteristics of the oceanic geoid and would require a well-calibrated, high accuracy inertial system. It has the potential to cover a range of the gravity field spectrum not well determined by current and future satellite missions.

6. CONCLUSIONS

A kinematic method to obtain relative geoid profiles from a combination of GPS and INS measurements has been studied with special emphasis on its sensitivity with respect to sensor accuracy, data density, and vehicle speed. The most important result is that requirements for inertial sensor accuracy and calibration are very high while modest GPS accuracies will be sufficient to achieve a relative geoid accuracy of better that 5 cm over distances of 50 km. Typical sensor accuracies are 0.001 degree per hour for gyro drift, a few mGals for accelerometer bias variation, about 10m for GPS position updates, and 0.1 m/s for velocity updates. The results are rather insensitive to changes in update data density and in gravity field variation. They are somewhat dependent on vehicle speed, although there is a rather large range between 40 km/h and 150 km/h where near optimal results can be achieved. Thus, land vehicle as well as airborne applications seem to be feasible. Some of these applications are briefly discussed in the paper.

It is interesting to note that the above conclusions do not hold for deflection determination. There is far more dependence on GPS data accuracy and density, as well as on gravity field variation and vehicle speed. This reflects the smoothing action of the integration and the corresponding insensitivity of the geoidal undulations to short wavelengths of the gravity field.

REFERENCES

Brown, R.G. (1983). *Introduction to random signal analysis and Kalman filtering.* John Wiley, New York.

Eisfeller, B. and P. Spietz (1989). Shaping filter design for the anomalous gravity field by means of spectral factorization. *Manuscripta Geodaetica*, 14, pp. 183-192.

Gelb, A. (ed.) (1974). *Applied optimal estimation.* M.I.T. Press, Cambridge, Mass.

Huddle, J.R. (1978). Theory and performance for position and gravity survey with an inertial system. *J. Guidance and Control*, Vol.I No.3, pp 183-188.

Jordan, S.K., P.J. Moonan, J.D. Weiss (1981). State space models of gravity disturbance gradients. In *IEEE Transactions on Aerospace and Electronic Systems*, Vol. AES-17, No. 5, Sept. 1981, pp. 610-618.

Kleusberg, A., A. Goodacre, R.J. Beach (1989). On the use of GPS for airborne gravimetry. *Proc. 5th Int. Geod. Symp. on Satellite Positioning*, Las Cruces, New Mexico, March 13-17.

Schwarz, K.P. (1983). Kalman filtering and optimal smoothing. In: E.J. Krakiwsky (ed.): *Papers for the CIS Adjustment and Analysis Seminars*, Calgary.

Schwarz, K.P. (1987). Geoid profiles from an integration of GPS satellite and inertial data. *Bolletino di geodesia e scienze affini*, N.2.

Schwarz, K.P. and M. Wei (1988). A framework for modelling kinematic measurements in gravity field applications. *Proc. AGU Chapman Conference on Progress in the Determination of the Earth's Gravity Field*, Ft. Lauderdale, September 12-16.

Vassiliou, A. and K.P. Schwarz (1987). Study of the high frequency spectrum of the anomalous gravity potential. *Journal of Geophysical Research*, Vol. 92, No. B1, pp. 609-617, January 10, 1987.

THE ITALIAN GRAVIMETRIC GEOID

B. Benciolini - Dip. Ingegneria - Università di Trento
A. Manzino - F. Sansò - Istituto di Topografia - Politecnico di Milano

Abstract
A new step of the Italian geoid has been almost completed. The geoid has been computed all over the Italian territory using a new gravity data base and a global model. The computation has been partly repeated using a digital height model too in order to separate the terrain effects; this job is still in development.
The data menagement and the computational activities are reported in this paper and the results are presented and discussed.

1. Introduction

This paper reports the work done and the results obtained at the Istituto di Topografia, Fotogrammetria e Geofisica del Politecnico di Milano in the last two years in the field of geoid computation.
Two main results have been achieved:
- the geoid has been computed all over Italy using a global gravity model in form of spherical harmonics and a set of gravity anomalies;
- the geoid has been pre-computed in a limited zone including a detailed DHM and testing various procedures for the evaluation of the terrain effect.

The first result is important because the geoid is now available all over Italy.
The second result is important because various procedures for separating the global model and the topographic contribution have been compared and tested; the gained experiences will be applied in the new computation of a more precise complete geoid witch is foreseen for the next year.

2. The data sets and their management

The original gravity data base has been prepared by the "Istituto di Fisica dell'Università di Lecce" and contains ~ 252.000 points. The data have been reordered and selected in order to obtain a more homogeneous set of point-value gravity anomalies partitioned into 22 zones. This has been obtained in several steps using the data base management programs developed at the institute.
The zones are designed in such a way that 11 of them completely cover the Italian territory, and the other 11 cover the borders of the basic ones. The DHM that we used has been prepared by the colleagues of Lecce too and consists of mean height values over 10"x7.5" blocks; the data are obtained from the existing maps at the scale 1:25.000.
These data has been organized in a data bank and a specific routine has been implemented to extract the data in a form suitable for subsequent computation.
The preprocessing of the DHM includes the selection of the data of a specific zone, completely independent by the original partition into sheets, and the computation of mean values over large blocks.

3. The estimation of the covariance function

The empirical covariance function has been computed separately for all the 22 zones, both for the raw data and the residual data after the removal of the contribution of a global model. The model GPM2 has been used for this purpose.
The empirical functions have been used to estimate the values of the parameters of a model covariance function (Tscherning, 1976), (Knudsen, 1987).
The results are summarized in the following.

	Δg^F			$\Delta g_{LOC} = \Delta g^F - \Delta g_{180}$					
zone	POW. $(mgal)^2$	C.L. (min.)	σn mgal	POW. $(mgal)^2$	C.L. (min.)	σn mgal	$R_B - R_E$ (Km)	a (N.S.)	n. points
Q_1	2860	13	18	1550	6	16	-1.2	(-0.33)	781
Q_2	2900	5	23	2790	7	22	-1.8	0.64	568
Q_3	2930	20	16	1970	17	5	-13.0	13	1483
Q_4	4580	17	15	3500	14	9	-8.0	7	1018
Q_5	3850	18	6	980	7	10	-1.3	(-1)	1366
Q_6	2840	23	8	780	10	9	-3.0	0.9	1127
Q_7	1300	7	11	1080	6	9	-0.9	(-0.4)	1376
Q_8	1110	13	10	930	10	10	-3.3	(-0.9)	1312
Q_9	2100	10	14	2500	12	12	-5.8	4.9	1286
Q_{10}	3500	21	4	3500	13	~0	-7.4	6.4	1278
Q_{11}	540	8	6	750	12	4	-3.1	1.1	723

The behaviour of the main parameters (i.e. the power of the signal and the correlation length) clearly indicates that the data are represented by the model with a very dishomogeneous accuracy; the agreement is particularly poor in the mountainous zones.
A more complete description of the data preparation and of the estimation of the covariance function can be found in (Benciolini e Manzino, 1987).

4. The main computation

The geoid has been separately computed in the already mentioned 22 zones, using the procedure of collocation applied to the local (i.e., residual w.r.t. the global model) data. The contribution of the global model has been subsequently restored in the result.

After the data preparation, that requested a significant human effort, the computation has been carried on as a routinary work, but it requested, on the other hand, an intensive use of a main-frame computer. The geoid has been computed on a grid with 0.125° spacing in both directions; the prediction error has been computed on a grid with 0.250° spacing in order to reduce computing time. This can be done because the prediction error of collocation is a quite smooth function and can be easily interpolated. The prediction error ranges from 0.10 m to 0.90 m.
The final result has been obtained by a very simple procedure, i.e. by taking the weighted mean of the values obtained for each point in different zones; the weights are the inverses of the variance of the prediction error.
The 22 zones where the geoid has been computed cover only the land; on the other hand it is quite convenient to have the result on a complete regular grid; this problem has been solved by merging the results of collocation and a grid of geoid undulations computed by a global model only.
The result is shown in fig. 1 (contour plot of the geoid) and in fig. 2 (contour plot of the estimation error).

5. The computation of the geoid including a DHM

The DHM can be used to remove the main local irregularities from the data in order to obtain a smoother field that can be better approximated by collocation. The effect of the local masses is removed from the data and restored in the result.
This procedure has been applied in a limited test zone.
The main difficulty of this procedure comes from the joint use of a global model (which is expected to remove a trend component from the data) and a DHM (which is expected to remove, via T.C., the very local components). In reality the T.C. computed considering all the masses between the geoid and the physical earth surface contains a long wave component. The overlapping of this component and the global model produces a systematic distortion in the result; a strategy is therefore needed in order to separate the contribution of the T.C. and the contribution of the global model.
The solution to this problem is the construction of a reference D.H.M. and the computation of the T.C. using the masses between the reference and the physical surfaces.
The reference D.H.M. is generally obtained by averaging the dense height data over large blocks; we used here this procedure and also a new one, based on a kind of geophysical inversion of the global gravity model.
The strong correlation of the global model and the trend of the topography is quite evident from fig. 3 (Δg from the GPM2) and fig. 4 (D.H.M.).
This procedure (Benciolini and Manzino, 1989) is essentially performed in two steps:

- the gravity anomalies Δg are evaluated on a grid, using the global model
- the gravity anomalies are pointwise transformed into an artificial topographic height using the Bouguer relation in a reverse way
  $h = \Delta g/a$.

These heights are then used as reference surface.

The coefficient a can be the standard one (a = 0.1119, if ρ = 2.67 g/cm³), or it can be estimated on a local base using the values of Δg and h ("true" value) and the Bouguer relation. The results obtained correspond to a density ρ = 2.36 using the L.S. criterion and ρ = 1.90 using the Least Absolute Residuals criterion.

The data have been analyzed after the application of 12 different treatments:

1) no treatment
2) only global model removed

3-12) global model and T.C. removed. The reference surface for the T.C. is:

3) h = 0 (complete correction)
4) h = 537 (tuned to obtain Δg = 0)
5) block average over 8x8 km²
6) " " " 16x16 "
7) " " " 32x32 "
8) " " " 64x64 "
9) from global model inversion using standard density (ρ = 2.67) and a vertical translation to obtain Δg = 0
10) from global model inversion using standard density
11) equal to 10, but ρ = 2.36
12) equal to 10, but ρ = 1.90

	Var. of the data	C_0 emp.	C_0 fit.	σ_n	C.L. (deg.)	R_E-R_B (Km)	(*)
1	2293	2294	1657	25.2	0.43	14.9	
2	2456	2458	2009	21.1	0.31	6.9	0.28
3	2474	2438	2296	13.3	0.31	23.3	0.20
4	2497	2460	2240	16.0	0.34	25.4	0.20
5	2101	2101	2058	6.5	0.32	8.1	0.25
6	1897	1938	2039	n.a.	0.32	8.1	0.25
7	1500	1477	1606	n.a.	0.39	27.4	0.20
8	730	736	724	2.5	0.37	38.7	0.15
9	979	950	728	15.9	0.32	20.1	0.20
10	885	878	841	6.7	0.35	24.5	0.18
11	849	843	829	3.4	0.32	23.6	0.17
12	892	889	872	4.4	0.34	20.4	0.18

(*) Mean value of the geoid prediction error

The previous table summarizes the main features of the covariance functions and the mean precision of the geoid.

The following remarks are in order:

- treatments 2÷7 are not able to significantly reduce the variance of the data (which is even increased in some zones)
- the noise is particularly high when the T.C. is not applied at all; this is due to the anisotropy of the topography and a sampling effect.
- the power of the signal is significantly reduced from case 7 to 8; this indicates that the proper filter has been found to separate the contributions of the global model and the T.C.
- cases 9÷12, based on the geophysical procedure, are similar to each other and also similar to case 8
- the shift of the reference height to obtain $\Delta g = 0$ does not contribute to smoothing the data (see 4 versus 3) and 9 versus 10).

The difference between the various solutions is still quite large (see fig. 5, 6) and is apparently not compatible with the prediction error which comes from collocation. It must be pointed out that the estimation error variance must be increased to at least 30 cm to take into account the effect of the noise (Benciolini and Manzino, 1989).

6. Comparison with some GPS-derived undulations

Some geoid differences have been obtained by the orthometric heights (from leveling) and ellipsoidal heights (from the GPS Campaign for the European Geotraverse).

The result are shown in the following table; it confirms what we already know, i.e. the need of using the T.C. to improve the precision of the geoid, through the final result is not yet fully satisfactory.

<table>
<tr><th>P.#</th><th>ϕ</th><th>λ</th><th>ζ_2</th><th>ζ_8</th><th>$\Delta\zeta_2$</th><th>$\Delta\zeta_8$</th></tr>
<tr><td rowspan="2">3</td><td rowspan="2">46.45469</td><td rowspan="2">11.33103</td><td rowspan="2">48.16</td><td rowspan="2">50.49</td><td></td><td></td></tr>
<tr><td rowspan="2">-0.48</td><td rowspan="2">-0.21</td></tr>
<tr><td rowspan="2">4</td><td rowspan="2">46.19008</td><td rowspan="2">11.13178</td><td rowspan="2">47.33</td><td rowspan="2">49.93</td></tr>
<tr><td rowspan="2">0.40</td><td rowspan="2">-0.11</td></tr>
<tr><td rowspan="2">5</td><td rowspan="2">45.90117</td><td rowspan="2">11.03611</td><td rowspan="2">47.40</td><td rowspan="2">49.49</td></tr>
<tr><td rowspan="2">0.83</td><td rowspan="2">-0.25</td></tr>
<tr><td rowspan="2">6</td><td rowspan="2">45.66833</td><td rowspan="2">10.90394</td><td rowspan="2">47.09</td><td rowspan="2">48.10</td></tr>
<tr><td></td><td></td></tr>
</table>

ζ = geoid estimated in cases 2 and 8 of § 5

$\Delta\zeta$ = discrepancies with the GPS geoid

7. Conclusions

A complete digital model of the geoid over Italy has been obtained; it will be used for various purposes including the (not high precision) conversion of ellipsoidal heights from GPS into orthometric heights.

The experience gained on the test zone will be applied to the computation of a more precise geoid all over Italy.

8. Acknowledgments

The computation of the Italian Geoid has been performed with the financial support of I.G.M.I.
Most of the applied main computing programs have been supplied by C.C. Tscherning and R. Forsberg.

9. References

Tscherning C.C., Rapp H.R. (1974): Closed covariance expressions for gravity anomalies, geoid undulations, and deflections of the vertical implied by anomaly degree variance models.
Reports n. 208. Dept. of Geodetic Science, The Ohio State University, Columbus.

Tscherning C.C. (1976): Covariance expression for second and lower order derivatives of the anomalous potential. Report n. 255, Dept. of Geodetic Science, The Ohio State University, Columbus.

Forsberg R., Tscherning C.C. (1981): The use of height data in gravity field approximation by collocation. Journal of Geophysical Research, vol. 86 n. B9, pp. 7843-7854.

Carrozzo M.T. et al. (1982): Data bases of mean height values and of gravity values. Proceedings of the 2nd International Symposium on the Geoid in Europe and Mediterranean Area, Roma.

Sideris M. (1985): A fast Fourier transform for computing terrain corrections. Manuscripta Geodaetica n. 10.

Barzaghi R. (1986): La stima del geoide in Italia. Ph.D. Thesis, Milano.

Knudsen P. (1987): Estimation and modelling of the local empirical covariance function using gravity and satellite altimeter data. Bullettin Geodesique, n. 61, pp. 145-160.

Benciolini B., Manzino A. (1987): Italgeo '90: La preparazione dei dati, la stima delle funzioni di covarianza ed i primi risultati ottenuti. Atti del 6¯ Convegno del GNGTS-CNR, Roma.

Gerstbach G. (1988): Precise Alpine geoid determination without digital terrain models. Bullettin Geodesique, n. 62, pp. 541-563.

Zomorrodian H. (1988): Regional geopotential model improvement for the Iranian geoid determination. Bullettin Geodesique, n. 62, pp. 125-141.

Benciolini B., Manzino A. (1989): The geoid computation: a fronteer between boundary value and inverse problems. II Hotine-Marussi Symposium on Mathematical Geodesy, Pisa, 1989.

FIG. 1

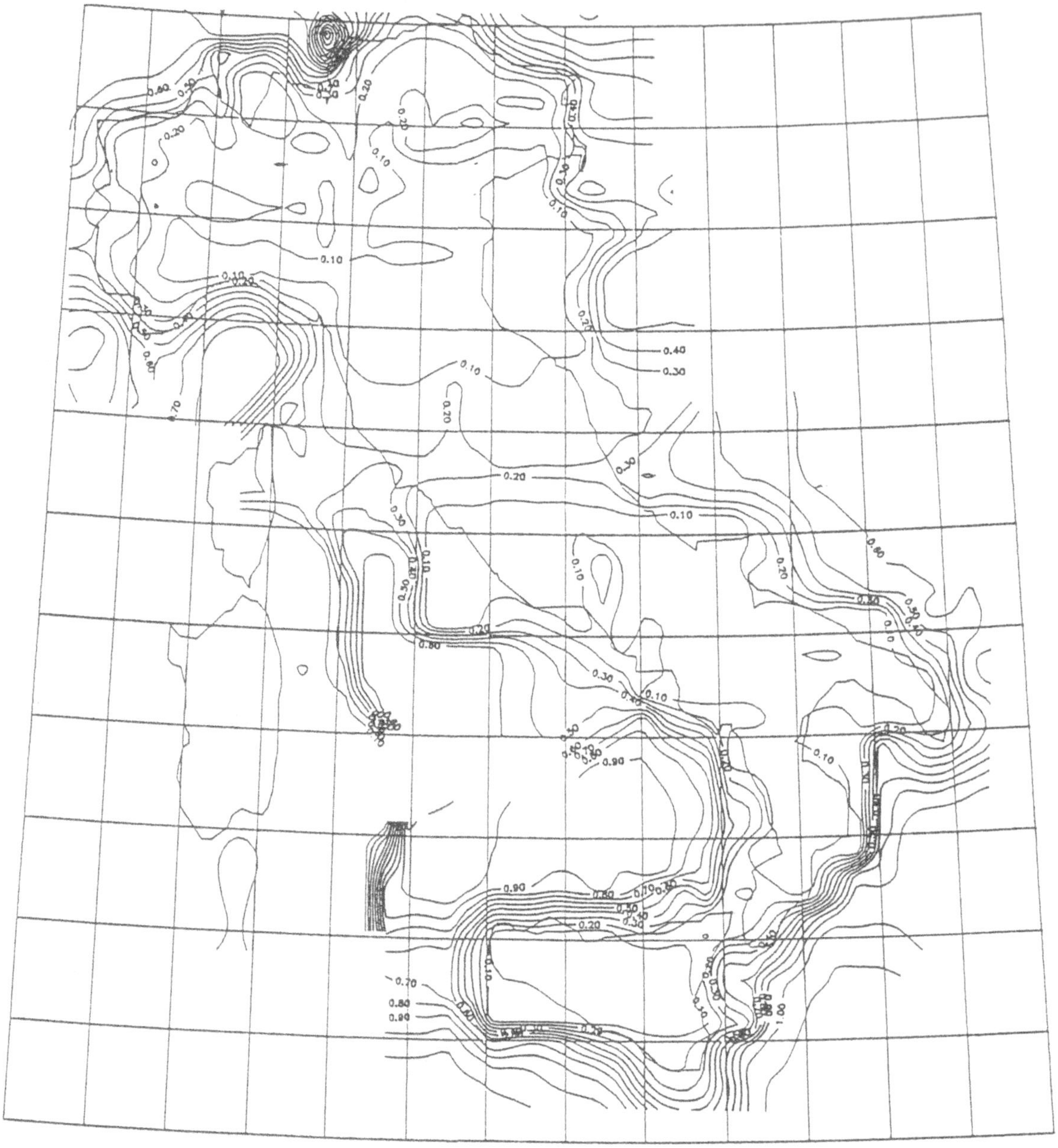
0.10
0.20
0.30
0.40
0.10
0.20
0.30
0.80
0.10
0.20
0.80
0.90
0.90
0.80
0.20
0.70
0.80
0.90
1.00

FIG. 2

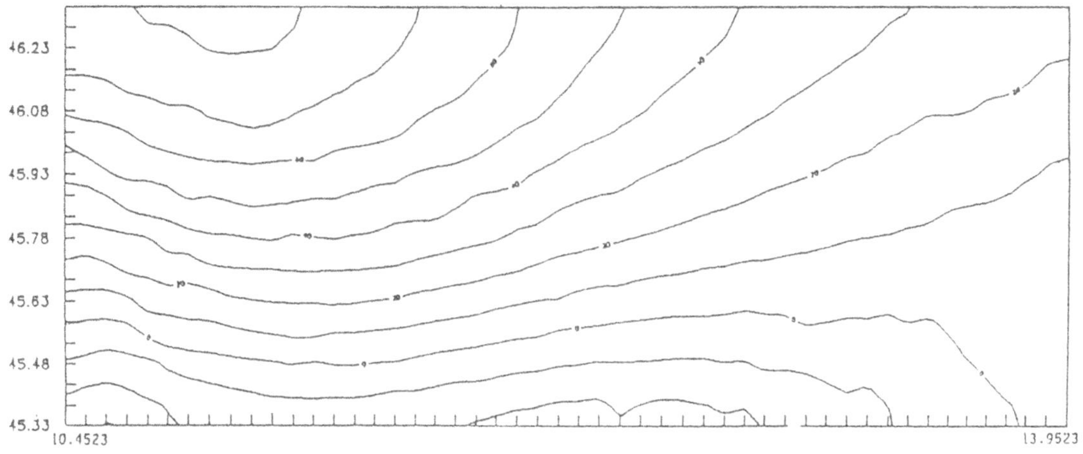
46.23
46.08
45.93
45.78
45.63
45.48
45.33
10.4523
13.9523

FIG. 3

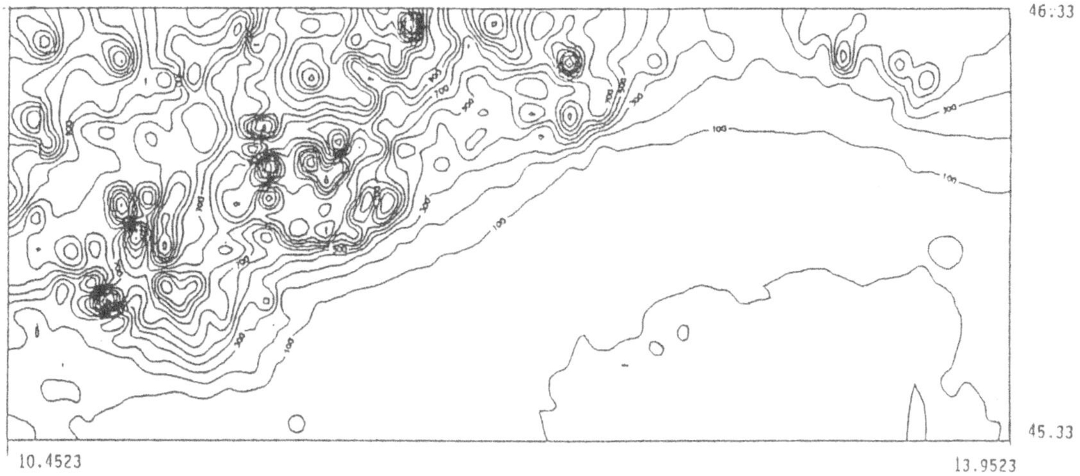
100
10.4523
13.9523
45.33

FIG. 4

64 KM GRID – PURE COLLOCATION (D = 0.5 m)

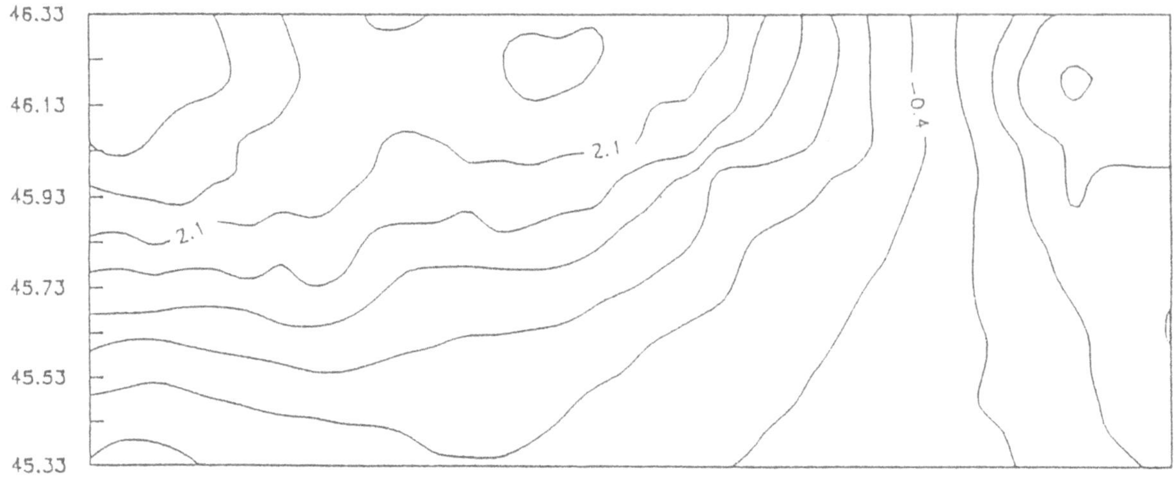

FIG. 5

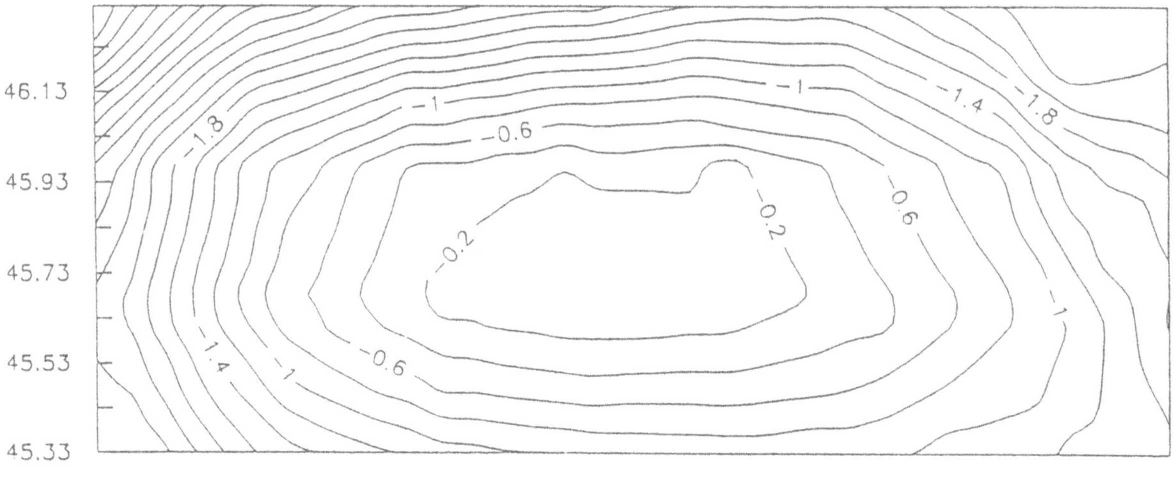

FIG. 6

KERNEL MODIFICATION IN GENERALIZED STOKES'S TECHNIQUE FOR GEOID DETERMINATION

Petr Vaníček
Department of Surveying Engineering
University of New Brunswick, Fredericton, N.B., Canada E3B 5A3

Lars E. Sjöberg
Department of Geodesy,
The Royal Institute of Technology, Stockholm, Sweden

This is a summary of two papers, "Reformulation of Stokes's Theory for Higher than Second-Degree Reference Field and Modification of Integration Kernels," and "The Modification of Stokes's Formula for Higher Degree Reference Field Considering Errors in Truncation, Harmonic Coefficients, and Terrestrial Data," submitted to the *Journal of Geophysical Research* by the authors.

INTRODUCTION

We wish to construct the geoid from a combination of satellite-determined potential coefficients (up to degree and order L) and terrestrial gravity anomalies Δg. We use the first $M \leq L$ degrees to define a reference spheroid. Terrestrial data are then included by means of Stokes-like convolution integration. Wishing to limit the integration area to a spherical cap of the smallest possible radius, we seek the most appropriate modified integration kernel.

REFERENCE SPHEROID

Of degree and order M:

$$\boxed{(N)_M = \sum_{i=2}^{M} N_i} \tag{1}$$

(preferably pure satellite dynamic solution $\Rightarrow M \sim 20$). Model gravity

$$\boxed{(\gamma)_M = \sum_{i=2}^{M} \gamma_i} \tag{2}$$

replaces γ_0 to give

$$\boxed{\Delta g^M = g - (\gamma)_M} \quad , \tag{3}$$

gravity anomaly referred to the reference spheroid ($\Rightarrow$ g assumed reduced to the geoid).

- Geocentric, properly oriented coordinate system is stipulated.
- As M grows, $\Delta g^M \to 0$.
- $\forall n > M : \Delta g_n^M = g_n$.
- Errors in potential coefficients induce errors in reference spheroid (mean square error now < 1 m) the same way as errors in "a" and "f" induce errors in the reference ellipsoid.

GENERALIZED STOKES'S FORMULA

Height of the geoid above the reference spheroid:

$$\boxed{N^M \doteq k \oiint_{\mathcal{E}} S^M(\psi)\, \Delta g^M \, d\mathcal{E}} \tag{4}$$

(spherical and other approximations at play), where

$k = (4\pi R \overline{\gamma})^{-1}$

and

$$S^M(\psi) = \sum_{n=M+1}^{\infty} \frac{2n+1}{n-1} P_n(\cos\psi)$$

is the spheroidal (generalized) Stokes's kernel.

- $(N)_M + N^M \doteq N$, as required.
- S^M tapers off more rapidly than S. As M grows, $\|S^M\| \to 0$ ($\forall M > 2 : \|S^M\| < \|S\|$):

$\Rightarrow$ (i) distant Δg^M have less effect on N;
(ii) approximations in Stokes's theory have less effect on N (for M=20 reduction by one order of magnitude).

- S^M is "blind" to first M degrees of Δg^M.
- Topographic, indirect, and atmospheric effects are left out of the discussion here.

TRUNCATED INTEGRATION

Equation (4) is rewritten as

$$\boxed{N^M \doteq k \iint_{\mathcal{C}_o} S^M(\psi)\, \Delta g^M \, d\mathcal{E} + k \iint_{\mathcal{E}-\mathcal{C}_o} S^M(\psi)\, \Delta g^M \, d\mathcal{E}}\,, \tag{5}$$

where the first term is the truncated integral, the second is the truncation correction (δN^M), and $\mathcal{C}_o$ is a spherical cap of radius ψ_o.

- For growing M, $\delta N^M \to 0$;
- For growing ψ_o, $\delta N^M \to 0$ (and $\mathcal{C}_o \to \mathcal{E}$);
- For $\psi_o \to 0$, δN^M grows to N^M;
- For practical reasons, we wish to have both ψ_o and $|\delta N^M|$ as small as possible $\Rightarrow$ compromise needed;
- Change (modification) of S^M affects the magnitude of $\delta N^M \to$ tool for optimization.

MOLODENSKIJ-LIKE MODIFICATION

Using all first L degrees, equation (5) gives

$$\boxed{N^M \doteq k \iint_{\mathcal{C}_o} S^{M,L}_{MOL}(\psi)\,\Delta g^M\, d\mathcal{E} + \delta N^{M,L}_{MOL}} \tag{6}$$

$$= N^{M,L}_{MOL} + \delta N^{M,L}_{MOL},$$

where

$$S^{M,L}_{MOL}(\psi) = S^M(\psi) - \sum_{n=2}^{L} \frac{2n+1}{2}\, t_n\, P_n(\cos\psi)\,.$$

Here, $t_1, t_2, \ldots, t_L$ are obtained from:

$$\forall n \le L: \sum_{k=2}^{L} \frac{2k+1}{2}\, e_{nk}(\psi)\, t_k = Q^L_n(\psi)\;,$$

where

$$Q^L_n(\psi) = Q_n(\psi) - \sum_{k=2}^{L} \frac{2k+1}{2}\, e_{nk}(\psi)\,,$$

$$Q_n(\psi_o) = \int_{\psi=\psi_o}^{\pi} S(\psi)\, P_n(\cos\psi)\, \sin\psi\, d\psi\,,$$

$$e_{nk}(\psi_o) = \int_{\psi=\psi_o}^{\pi} P_n(\cos\psi)\, P_k(\cos\psi)\, \sin\psi\, d\psi\,.$$

In spectral (series) form:

$$\boxed{N^{M,L}_{MOL} \doteq -C \sum_{n=2}^{L} t_n\, \Delta g^M_n + C \sum_{n=L+1}^{\infty} \left(\frac{2}{n-1} - Q^{L*}_n\right) g_n}\,, \tag{7}$$

$$\boxed{\delta N^{M,L}_{MOL} \doteq C \sum_{n=2}^{L} \left(\frac{2}{n-1} + t_n\right) \Delta g^M_n + C \sum_{n=L+1}^{\infty} Q^{L*}_n\, g_n}\,, \tag{8}$$

where

$$C = \frac{R}{2\gamma}\,,$$

$$\forall n: Q^{L*}_n = Q^L_n - \sum_{k=2}^{L} \frac{2k+1}{2}\, e_{kn}\, t_k\;.$$

- This modification minimizes the upper bound of $|\delta^{M,L}|$;
- As $\|\Delta g^M\| < \|\Delta g\|$, $|\delta N^{M,L}_{MOL}| < |\delta N^{\bullet,L}_{MOL}|$, with $\delta N^{\bullet,L}_{MOL}$ denoting the correction for the standard Molodenskij modification.

⇒ (i) for the same ψ_o, truncation error is smaller for the generalized Stokes approach;
(ii) for the same truncation error, ψ_o is smaller for the generalized case;

- This technique was used (for L=M) to compile the University of New Brunswick "UNB Dec.'86" geoid for Canada (errors smaller than 2×10^{-6} in non-mountainous regions).

OBSERVATIONAL ERRORS

Denote
ε^T ... error in "observed" g;
ε^S ... error in model field γ generated by potential coefficients up to degree L;
ε_M ... error in $(\gamma)_M$;
$\forall n \le L : \varepsilon_n^S$ ... error in γ_n ($\Rightarrow \varepsilon_M = \sum_{n=2}^{M} \varepsilon_n^S$, $\varepsilon^S = \sum_{n=2}^{L} \varepsilon_n^S$);

$\forall n : \varepsilon_n^T$ ... error in g_n.

For "observed" Δg^M:

$$\boxed{\hat{\Delta} g^M = \Delta g^M + \varepsilon^T - \varepsilon_M} \tag{9}$$

$$\boxed{\Rightarrow \hat{\Delta} g_n^M = \begin{cases} \varepsilon_n^T - \varepsilon_n^S & \forall n \le M \\ g_n + \varepsilon_n^T & \forall n > M \end{cases}} . \tag{10}$$

Assume:

$\forall n : E(\varepsilon_n^T) = 0, \forall n \le L : E(\varepsilon_n^S) = 0$

$$\boxed{\Rightarrow E(\varepsilon^T) = E(\varepsilon^S) = E(\varepsilon^M) = 0} , \tag{11}$$

no local bias in g, or γ, or $(\gamma)_M$.
Denote:

$\forall n,k \le L : E(\varepsilon_n^T \varepsilon_k^T) = \Lambda_{nk}$,

$\forall n,k : E(\varepsilon_n^S \varepsilon_k^S) = \Omega_{nk}$.

- These covariances are position dependent.

Assume:

$\forall n,k : E(\varepsilon_n^S \varepsilon_n^T) = 0$.

- Assumes that terrestrial gravity is <u>not</u> used in the evaluation of potential coefficients.

Denote:

$$\Gamma(\bullet) = \frac{1}{\mathcal{E}} \oint\!\!\oint_{\mathcal{E}} \bullet \, d\mathcal{E} .$$

We get

$$\boxed{\Gamma(\Lambda_{nk}) = \begin{cases} (\sigma_n^T)^2 & \forall n = k \\ 0 & \forall n \neq k \end{cases}} \quad . \tag{12}$$

$$\boxed{\Gamma(\Omega_{nk}) = \begin{cases} (\sigma_n^S)^2 & \forall n = k \\ 0 & \forall n \neq k \end{cases}} \quad . \tag{13}$$

where $(\sigma_n^T)^2$, $(\sigma_n^S)^2$ are "error degree variances."

- Equations (12) and (13) follow from global orthogonality;
- Error degree variances depend on the data used.

FIELD VARIATIONS

Denote:

$$\Delta g_n = \begin{cases} \Delta g_n^M & \forall n \leq M \\ g_n & \forall n \geq M \, ; \end{cases}$$

$$\boxed{\Rightarrow \ \forall n : \Gamma(\Delta g_n) = 0} \ . \tag{14}$$

Denote:

$$\boxed{\forall n : c_n = \Gamma(\Delta g_n^2)} \ ,$$

the "field degree variances."

- These have nothing to do with observing errors or the data used.

We get

$$\boxed{\forall n \neq k : \Gamma(\Delta g_n, \Delta g_k) = 0}$$

from global orthogonality.

TWO GENERIC MODIFICATIONS

The two following schemes are generic:

$$\boxed{\tilde{N}^{M,L} \doteq k \iint_{\mathcal{C}_o} S^{\bullet,L}(\psi) \, \Delta g^M \, d\mathcal{E} + C \sum_{n=M+1}^{L} (Q_n^L + s_n) \, \gamma_n} \tag{15}$$

$$\boxed{\tilde{\tilde{N}}^{M,L} \doteq k \iint_{\mathcal{C}_o} S^{\bullet,L}(\psi) \, \Delta g^M \, d\mathcal{E} + C \sum_{n=M+1}^{L} s_n \, \gamma_n} \ , \tag{16}$$

where

$$S^{\bullet,L}(\psi) = S(\psi) - \sum_{k=2}^{L} \frac{2k+1}{2} s_n P_n(\cos\psi)$$

and the "modification coefficients" s_n are unspecified.

- For Molodenskij-like modification — $\forall n \leq L$: $s_n = t_n$, $Q_n^L = 0$

$\Rightarrow \tilde{N}^{M,L}_{MOL} = \tilde{\tilde{N}}^{M,L}_{MOL} = N^{M,L}_{MOL}$.

Total (including truncation) errors in the two generic schemes are: **Local bias**

$$\boxed{E(\varepsilon(\tilde{N})) = - C \sum_{n=L+1}^{\infty} Q_n^L g_n \neq 0}, \tag{17}$$

$$\boxed{E(\varepsilon(\tilde{\tilde{N}})) = - C \sum_{n=M+1}^{\infty} Q_n^L g_n \neq 0} . \tag{18}$$

- Both modifications are locally biased (bias changes with location).

Global Bias

$$\boxed{\Gamma(E(\varepsilon(\tilde{N}))) = - C \sum_{n=L+1}^{\infty} Q_n^L \Gamma(g_n) = 0}, \tag{19}$$

$$\boxed{\Gamma(E(\varepsilon(\tilde{\tilde{N}}))) = - C \sum_{n=M+1}^{\infty} Q_n^L \Gamma(g_n) = 0} . \tag{20}$$

- Both modifications are globally unbiased.

Defining

$MSE(\bullet) = VAR(\bullet) + BIAS^2(\bullet)$

we get: **Local MSE**

$$\boxed{MSE(\tilde{N}) = C^2 \Bigg[\sum_{n=2}^{\infty} \left(\frac{2}{n-1} - s_n^* - Q_n^L \right) \sum_{k=2}^{\infty} \left(\frac{2}{k-1} - s_k^* - Q_k^L \right) \Lambda_{nk}}$$

$$\boxed{+ \sum_{n=2}^{L} (Q_n^L + s_n) \sum_{k=2}^{L} (Q_k^L + s_k) \, \Omega_{nk} + \left(\sum_{n=L+1}^{\infty} Q_n^L g_n \right)^2 \Bigg]}, \tag{21}$$

where

$$\boxed{s_n^* = \begin{cases} s_n & \forall n \leq L \\ 0 & \forall n > L \end{cases}} .$$

Similar expression holds for $\tilde{\tilde{N}}$.

- MSEs are position dependent because covariances are position dependent.
- Any optimization has to change from location to location — impractical.

More modest: **Global MSE**

$$\Gamma(MSE(\tilde{N}))=C^2\left[\sum_{n=2}^{\infty}\left(\frac{2}{n-1}-s_n^*-Q_n^L\right)^2(\sigma_n^T)^2+\sum_{n=2}^{L}(Q_n^L+s_n)^2(\sigma_n^S)^2+\sum_{n=L+1}^{\infty}(Q_n^L)^2\,C_n\right], \quad (22)$$

$$\Gamma(MSE(\tilde{\tilde{N}}))=C^2\left[\sum_{n=2}^{\infty}\left(\frac{2}{n-1}-s_n^*-Q_n^L\right)^2(\sigma_n^T)^2+\sum_{n=2}^{L}(s_n^{**})^2\,(\sigma_n^S)^2+\sum_{n=M+1}^{\infty}(Q_n^L)^2C_n\right], \quad (23)$$

where

$$s_n^{**}=\begin{cases} Q_n^L+s_n & \forall n \leq M \\ s_n & \forall M < n \leq L \end{cases}.$$

- $\Gamma(MSE(\tilde{N}))$ is independent of M.

Equations (22) and (23) provide the tool for optimizing the modification by choosing appropriate s_n's.

THREE EXAMPLES

(i) Molodenskij's choice $\forall n \leq L : s_n = t_n$ in either generic scheme gives

$$\Gamma(MSE(N_{MOL})) = C^2\left\{\sum_{n=2}^{L}\left[\left(\frac{2}{n-1}-t_n\right)^2(\sigma_n^T)^2+t_n^2\,(\sigma_n^S)^2\right]\right.$$
$$\left.+\sum_{n=L+1}^{\infty}\left[\left(\frac{2}{n-1}-Q_n^L\right)^2(\sigma_n^T)^2+(Q_n^L)^2\,C_n\right]\right\}. \quad (24)$$

- Independent of choice of M.

(ii) Selection $\forall n \leq L : s_n = \frac{2}{n-1} - Q_n^L$ in the first generic scheme gives

$$\Gamma(MSE(N))=C^2\left\{\sum_{n=2}^{L}\left(\frac{2}{n-1}\right)^2(\sigma_n^S)^2+\sum_{n=L+1}^{\infty}\left[\left(\frac{2}{n-1}-Q_n^L\right)^2(\sigma_n^T)^2+(Q_n^L)^2\,C_n\right]\right\}. \quad (25)$$

- Observational errors in potential coefficients and terrestrial gravity contributions are strictly separated into two frequency bands.

(iii) "Least-squares modification" minimizes the global MSE. (Minimization of local MSE possible but impractical.) It leads to the following choice:

$$\underline{s} = \hat{\underline{s}} = \underline{A}^{-1}\,\underline{h},$$

where $\underline{A}$ is a regular symmetrical matrix and $\underline{h}$ is a vector, both constructed from field and error degree variances.

- Two versions are obtained from the two generic schemes, locally biased from L or M up respectively.

CONCLUSIONS

- Reference spheroid defined by potential coefficients determined from satellite dynamics should be used. Provides transparency of results; reduces long wavelength errors in the theory.
- Least-squares modification is the most desirable. But it requires tailoring to the data sets used — impractical.
- Locally best least-squares modification would be desirable but it is impractical — integration kernel changes from point to point.
- Globally best least-squares modification for some typical error spectra should be designed.

A HIGH RESOLUTION, HIGH PRECISION GEOID FOR THE BRITISH ISLES

M.P.Stewart, R.G.Hipkin

Department of Geology and Geophysics

University of Edinburgh, Edinburgh, EH9 3JZ, UK

1. Introduction

This paper demonstrates how the height of the geoid above the reference ellipsoid can be computed from surface gravity observations with high local precision. The geoid is the equipotential surface, closely approximated by mean sea level, which forms the zero-reference for conventional measurements of topographic height. The Global Positioning System (GPS) can measure the height of a ground-based receiver above the reference ellipsoid with a precision of a few centimetres. Consequently, a combination of GPS measurements with a precise gravimetric geoid can provide a cheap alternative to nationwide geodetic levelling.

The most detailed high resolution global gravity models currently available eg OSU86E and OSU86F (Rapp and Cruz 1987), are capable of resolving all wavelengths greater than about 110km. In a detailed study of a region of rough terrain and complex geology, Hipkin (1988) found 1m high geoid features with peak-to-trough distances of about 30 km. Such features are well beyond the typically 100km resolution sought by projected satellite missions using satellite-to-satellite tracking or gradiometry. The highest resolution models will remain dependent upon surface integration of gravity anomalies. These local methods of determining a gravimetric geoid are all essentially derived from Stokes' integral.

Hipkin (1988) described a modification of Stokes' method, capable of yielding a very high resolution geoid. It differed from the classical approach in two ways. First, it investigated a generally unreported proposal by Stokes to integrate separately the Bouguer anomaly and the attraction of a terrain model. Secondly, it made use of the Fast Fourier Transform (FFT) algorithm by assuming that a flat Earth approximation was applicable to the size of the data rectangle. This method generates a local geoid relative to an unknown plane, rather than an 'absolute' geoid relative to a known ellipsoidal reference surface. Here we demonstrate one method of obtaining an 'absolute' geoid using the FFT method. Rather than detrending gravity data with a least squares plane, we now subtract a high degree and order spherical harmonic gravity model.

2. Theory

On land the deviation potential can be split into 2 parts

$$V = V_T + V_B \tag{1}$$

where V_T is the potential due to the visible topography condensed to a surface density at zero height and V_B is the potential due to anomalous densities within and below the topographic masses (Bouguer anomaly potential). At sea, V_T is set equal to zero so that V_B is identical to the deviation potential.

Geoid undulations for each part, N_T and N_B, can be calculated with adequate accuracy by representing Stokes' integral in the frequency domain.

$$N_B = \frac{1}{\gamma} F^{-1}\left[\frac{F[\Delta g]}{k} \right] \tag{2}$$

where

γ = normal gravity
k = wavenumber (2π/wavelength)
G = gravity constant
ρ = constant defined density (2700 kg.m^{-3})
h = topographic height
Δg = Bouguer gravity anomaly on land and free air anomaly at sea
$F[\ ]$ denotes fourier transform
$F^{-1}[\ ]$ denotes inverse fourier transform

In order to minimise spurious contamination spreading from marginal discontinuities, it is necessary to suppress any long wavelength trends present in the data. The long wavelength component of the gravity potential can be calculated on the ellipsoid using a high degree global gravity model. This is used to remove long wavelengths prior to computation and is later restored. The long wavelength component is computed from

$$V_{GM} = \frac{GM}{R(\theta)} \sum_{n=2}^{nmax} \sum_{m=0}^{n} \left[\frac{a}{R(\theta)}\right]^n (C_{nm}(a)\cos m\lambda + S_{nm}(a)\sin m\lambda) P_{nm}(\cos\theta) \tag{3}$$

where

$R(\theta)$ = radial distance to computation point on the ellipsoid
nmax = maximum degree of spherical harmonic expansion
C_{nm}, S_{nm} = normalised coefficients of the anomalous potential
$P_{nm}(\cos\theta)$ = Legendre polynomials
λ, θ = geocentric longitude and colatitude
a = equatorial radius of the Earth

3. Data

3.1 Global Gravity Models

Three spherical harmonic solutions for the Earth's gravity field were used in this study: OSU86E and OSU86F (Rapp and Cruz 1986) are complete to degree and order 360 (the latter includes geophysically predicted anomalies in parts of the world where real surface gravity data are not available) and GPM2 (Wenzel, 1985) which is complete to degree and order 200. Free air gravity anomalies and geoid heights were generated on the GRS80 ellipsoid (equatorial radius 6378137m, reciprocal flattening 298.257 222) on a 20' x 15' grid and then interpolated to a 2km grid within a 900km x 1000km region of the British National Grid.

3.2 Gravity Data

The bulk of the gravity data are derived from the Edinburgh Gravity Data Bank which was used previously to produce the Bouguer anomaly map of the British Isles (Hipkin et al. 1986). Within this region the data bank stores some 400000 point observations including British national grid coordinates , topographic height, free air anomaly and Bouguer anomaly for each station. For this study the gravity data set was extended eastwards across the North Sea so that the Lowestoft tide gauge, the most easterly point of an associated test network of GPS observations (Ashkenazi et al. 1988), was not too close to the data margin. On the British mainland we used the same 2km grid of Bouguer anomalies from which the published Bouguer anomaly map was generated. At sea and where no adequate topographic model was available (Eastern Ireland, Northern France, Belgium and the Netherlands) a new data set of 2km gridded free air anomalies was generated from point observations in the data bank. Additionally, 2880 mean free air anomalies (blocksize 3' x 5') were included from the Netherlands, provided by Dr. G. Strang van Hees (pers comm. 1986). These observations were interpolated onto a 2km grid on land in the Low Countries. Their inclusion results in the rectangular data set being incomplete in the extreme southeast of the study area, that is in parts of Belgium, northeastern France and southern Norway. Here, free air anomalies as calculated from the spherical harmonic coefficents (eq. 3) were used. All interpolated point data have been uniformly reduced to GRS80. A uniform Bouguer and normal terrain density of 2700 $kg.m^{-3}$ was used throughout.

3.3 Topographic Data

The data used for the digital terrain model was derived from two sources . The first was a series of 1km mean elevations provided by the British Geological Survey (I.Smith, pers comm 1988). This covers most of Wales, SW England, the Southern Pennines, the Lake District and parts of East Anglia. For the rest of the country point heights associated with gravity stations in the gravity data bank were interpolated onto a regular grid. Although these do not represent mean heights the regions covered are of sufficiently low relief that point heights provide an adequate model for the topography. The model was set to zero in regions where the gravity data consisted of free air anomalies. The two forms of data were combined and interpolated onto the same 2km grid as the gravity anomalies.

3.4 Geoid Solutions

Because calculating an FFT of array dimensions 450x500 exceeds the capacity of the mainframe computer at Edinburgh, the data set was split into 4 overlapping 700km x 700km segments (figure 1). 2km gridded free air anomalies calculated from the global potential model OSU86E were subtracted from the array of local gravity anomalies. The resulting residual anomalies were then transformed to the gravity related component of the relative geoid (eq. 2). This was added to the component due to the topographic model to give the local geoid relative to OSU86E (figure 2). The absolute geoid N360E, shown in figure 3 with respect to the GRS80 spheroid, is obtained by adding the relative local components to the long wavelength component derived directly from OSU86E. This process was repeated for both other global models, OSU86F and GPM2 to give geoid solutions N360F and N200G.

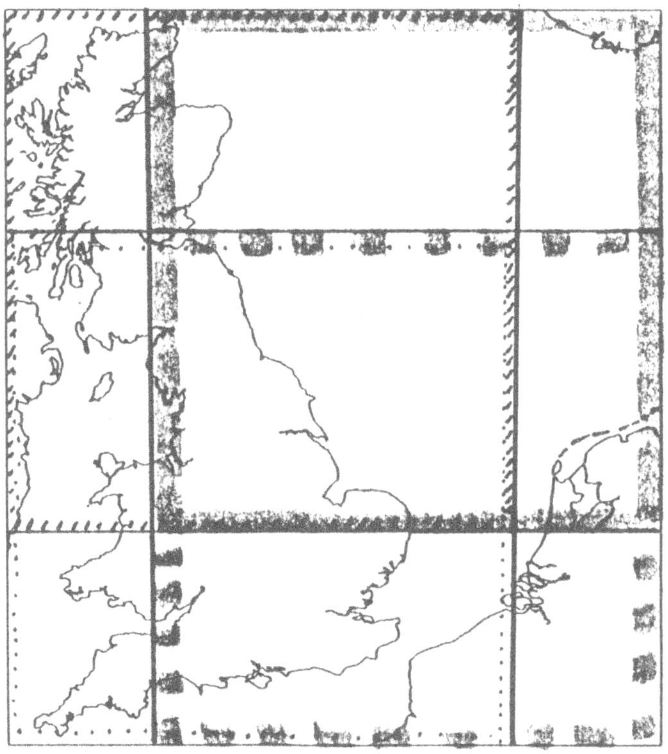

Fig 1: Overlapping regions used for FFT Computation

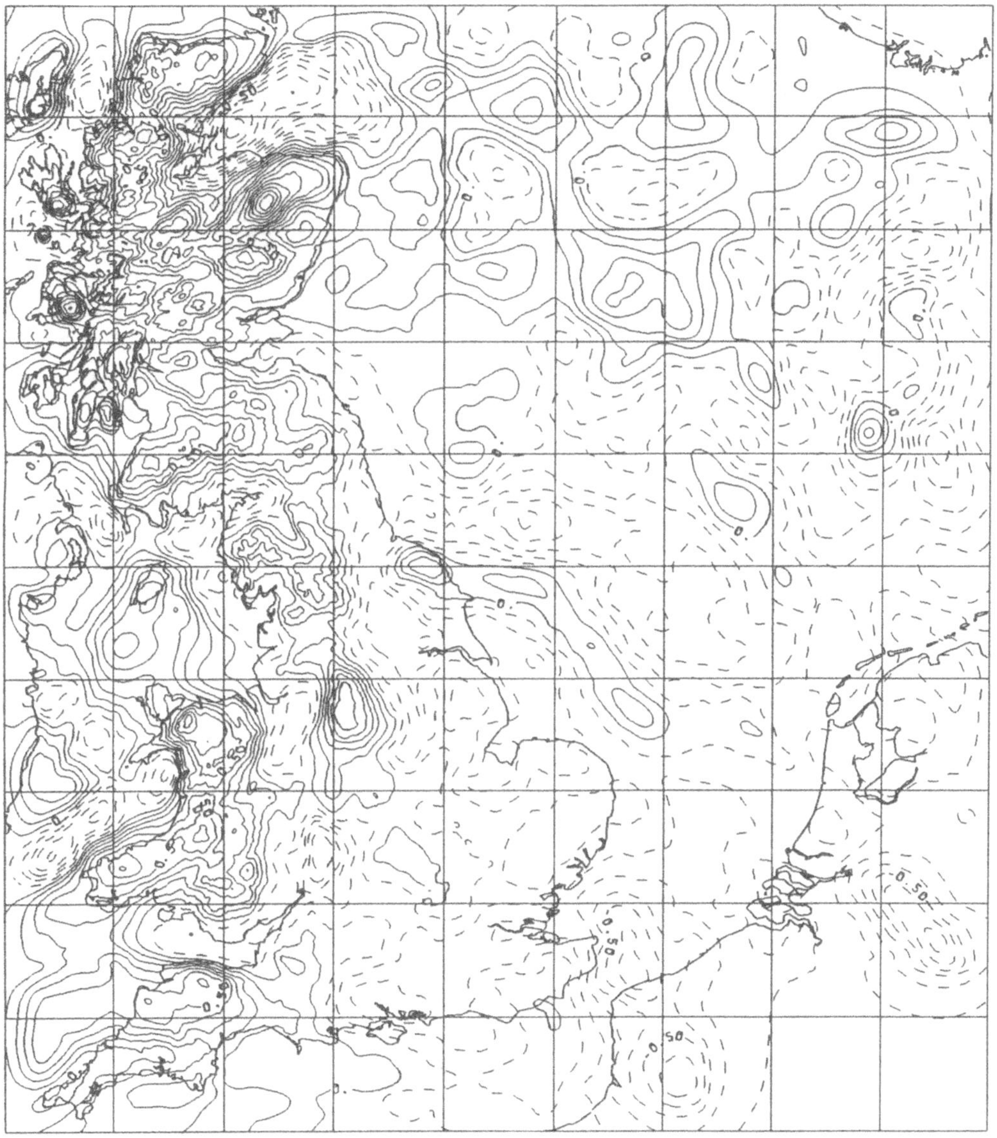

Fig 2: Detrended Geoid Relative to OSU86E
(0.10m contours, dashed contours negative)

4. Comparison of computed geoids

4.1 Long Wavelength Discrepancies between Global Geopotential Models

Our FFT method provides no constraint on the regional mean height of the geoid, nor on local trends which result from wavelengths longer than the sides of the data rectangle. Consequently, any very long wavelength errors in the spherical harmonic model, particularly in its datum, will remain in the finally computed geoid. This can be seen most clearly from deviations between geoid heights derived from solutions using different initial geopotential models. A least squares plane fitted to the height differences N360E-N360F gives a mean offset of -0.18m and a slope of 0.40m per 1000km increasing toward a bearing of N262° E. The corresponding results for N200G-N360E are a mean offset of 0.34m and a slope of 0.91m per 1000km towards N76° E. For these quantitive comparisons, a 50km wide strip around the edges of the data set was excluded because of leakage generated by the Fourier transformation. For the remaining 800km x 900km region the least squares plane generates 98% of the variance of N360F - N360E and 88% of the variance of N360E - N200.

The linearity of the deviations over an area of this size suggests that there are discrepancies between the different global gravity models of several decimetres in geoid height components with wavelengths longer than 2000 to 3000 km. No one data source provides the dominant control for these gravity models in this wavelength range (spherical harmonic degrees < 20), although the range is included in satellite-only models such as GRIML1 and GEMT1. The weakness of satellite-only solutions in this range is clearly demonstrated by the large change in the degree variances for potential GEMT1 solutions when a priori constraints such as Kaula's Law are included, and by the degree variance differences between GEMT1 and models including satellite altimeter data such as PGSS1(Marsh et al, 1988).

4.2 Short and medium wavelength discrepanices

Significant errors were found in areas of overlap between adjacent geoid solutions. Figure 4 shows the 2D error for a typical overlap region for geoid solution N360E. Noticeable is the effect of marginal leakage but longer wavelength features, in the order of 0-10cm, occur well away from the edges of the data set. These features are caused by the inability of the FFT algorithm to cope with wavelengths longer than about 1/3 the width of the data set which are still present after detrending. It was found that these errors could be minimised by controlling the shape of the data set and the extent to which the edges of the data were tapered. Significantly, figure 4 shows that short wavelength components match almost perfectly. No distortion of the short wavelength portion of the spectrum occurs using the FFT method. In comparing geoids computed directly from OSU86E, OSU86F and GPM2 and our computed geoids it is also apparent that the FFT method has the ability to improve intermediate wavelength components of the geoid. We have found that truncating the high degree and order spherical harmonic expansions at degree and order 200 has no effect on our final geoid. In this wavelength range local gravity data is dominant,

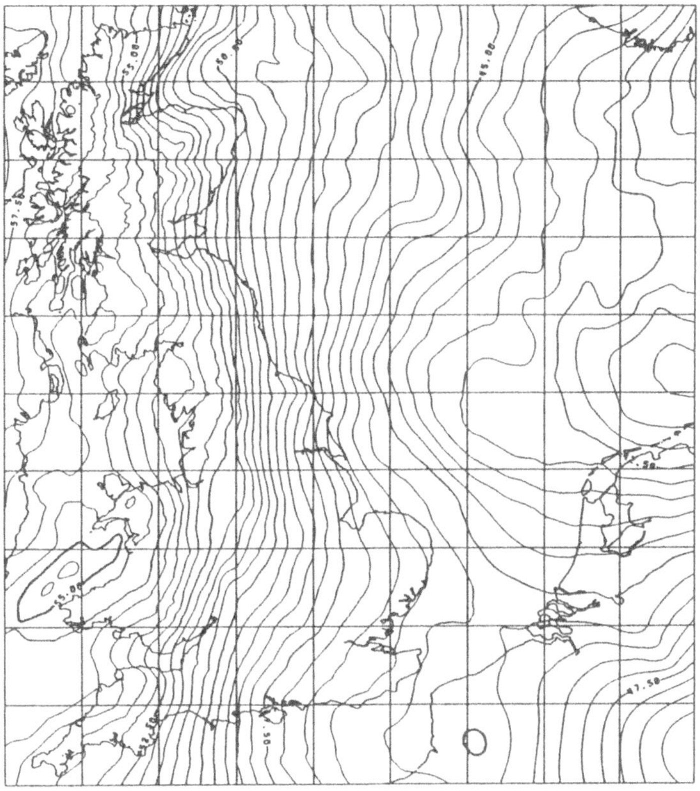

Fig 3: Absolute geoid, retrended using OSU86E (0.5m contours)

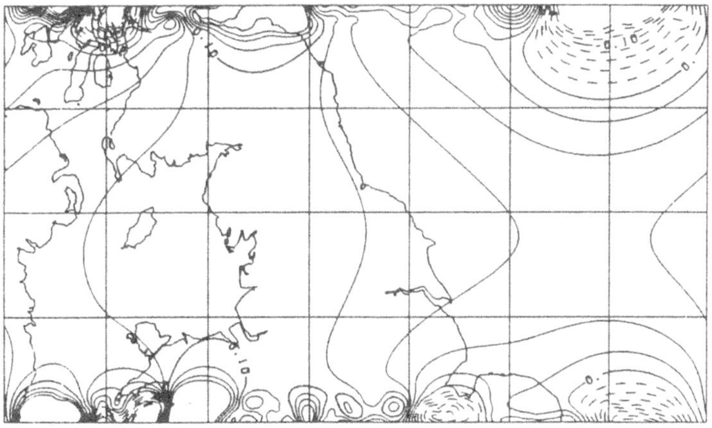

Fig 4: Typical error between adjacent overlapping segments (0.02m contours)

making geopotential coefficients and their associated errors redundant.

5. Summary

- In the 4 - 300km wavelength range, computed geoids differ at the 1cm level.
- Between wavelengths 300-1000km we see differences of up to 10cm caused by discrepancies between the geopotential model and the local gravity data resulting in residual long wavelength trends remaining in the gravity data prior to computation.
- For wavelengths greater than the dimensions of the data rectangle very long wavelength discrepancies between geopotential models cause differences between absolute geoids in the order of several decimetres.

Final results, including comparisons between FFT solutions, geoid heights derived from GPS observations during the Nottingham - Edinburgh Tide Gauge Project and a ring integration method (Kearsley, 1986), are currently in preparation.

6. References

Ashkenazi V et al. (1988) *Levelling of tide gauges using GPS*, Inst. of Eng. Surv. and Space Geod., University of Nottingham

Hipkin, R.G., D. Lyness, B. Chacksfield, C. Reay, D. Gibert, A.G. Turnbull, G.I. Lumsden (1988) *Bouguer Anomaly Map of the British Isles*, Brit. Geol. Surv. Nottingham.

Hipkin, R.G. (1988) Bouguer anomalies and the Geoid:a reassessment of Stokes' method. *Geophys. J. R. Astron. Soc.*,**92**, 53-66.

Kearsley, A.W.H. (1986) The determination of precise geoid differences using ring integration. *Boll. Geod. Sci.Affini, XLV(2)*, 151 - 174.

Marsh, J.G.,F.J. Lerch, B.H. Putney, D.C Christodoulidis, D.E. Smith, T.L. Felsentreger and B.V. Sanchez (1988) A new gravitational model for the Earth from satellite tracking data:GEM-T1, *J. Geophy. Res.*,**93**,6169-6215.

Rapp R. and J. Cruz (1986) Spherical Harmonic Expansions of the Earth's Gravitational Potential to Degree 360 using 30' Mean Anomalies, *Dept. of Geod. Sci. and Surv.Rep.* **376**, The Ohio State Univ., Columbus.

Wenzel,H.G (1985) Hochauflösende Kugelfunktionsmodelle für das Gravitationspotential der Erde, *Wiss. Arb. der Vermessungswesen der Univ. Hannover.* **137**, Hannover/FRG

A RELATIVE GEOID FOR THE UK

A H Dodson and S Gerrard
Institute of Engineering Surveying and Space Geodesy
Nottingham University, England

INTRODUCTION

Great Britain is in need of a precise relative geoid, primarily to enable the use of the Global Positioning System (GPS) for levelling. In 1987 a collaborative project was started between the University of Nottingham, the Ordnance Survey of Great Britain (OS) and the British Geological Survey (BGS) that would result in a geoid database covering the whole country. This database would be tried and tested as part of the project to ensure its suitability for practical use in Engineering and Topographic Surveying. This implied that geoid height differences should be accurate to around 1 cm, particularly over lines of up to 50 km, to be in agreement with third, or even second order levelling tolerances at 12 mm and 6 mm $\sqrt{km}$ respectively.

The method of Least Squares Collocation (LSC) was chosen for the determination of the British Geoid - a well documented technique that has already been used in other countries with some success, (Arabelos, 1989 and Benciolini et al, 1983). The technique would be implemented using software acquired from the Kort-og Matrikelstyrelsen, Denmark.

There are several key decisions to be made when implementing an LSC solution, including the choice of Geopotential Model, determination of a suitable covariance function and data selection. Much background work has been carried out on these factors but for the purpose of this paper only that concerning three areas will be discussed.

The Three Schemes

Three GPS observation Schemes were undertaken by the OS in April 1989, in order to provide 'observed' geoid height differences, ΔN^o, as a check on those predicted, ΔN^p. The areas were chosen specifically to represent the range of topography in Great Britain: North Wales (NW), by British standards mountainous, Derbyshire (DY), which is hilly but smooth, and East Anglia (EA) being basically flat. Each scheme involved eight stations based on OS Levelling marks, covering a range of distances (7 - 65 km) and directions, as can be seen in Fig 1.

Gravity data acquired from the BGS, was selected for each scheme covering an area of approximately 1.2° square, with a data point every 5 km or so (2.5'Ø x 4.0 'λ), giving around 600 points per area, (Fig 2). The gravity values were converted to free-air anomalies on GRS 80.

Two high degree Geopotential models were tested for each area to find the closest agreement to the geoid. These are OSU86E (Ohio State University) and IFE88E2 (Institut für Erdmessung, Hanover), both determined to degree and order 360, for a spatial geoid resolution of 1/2°. The two show a similar geoid trend of an East-West slope across the country, but differ by over a metre in places (Fig 3.)

• Height Station

Fig 1 The three schemes

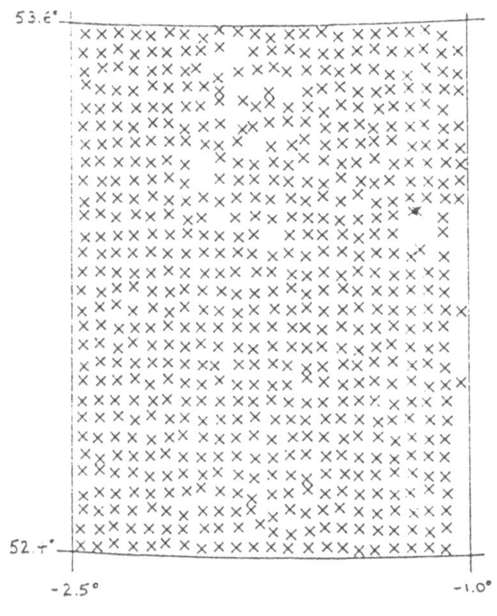

X Data point

Fig 2 Typical data distribution

It was therefore necessary to decide, using the three areas, which model describes the geoid most closely in Great Britain.

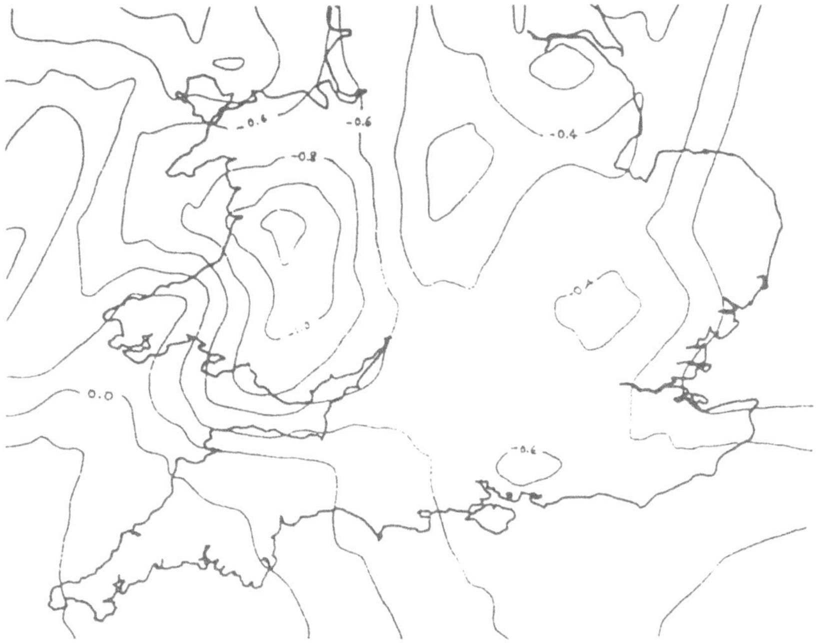

Fig 3 Residual Geoid: IFE88E2 - OSU86E, Contour interval = 0.2 m

Covariance Functions

The three data sets were reduced by each model such that empirical covariance functions could be determined, to describe the reduced gravity field characteristics. The associated parameters for these functions are seen in Table 1, and it is apparent from the mean values and standard deviations that OSU86E reduces the data to a greater extent in most cases. Also in East Anglia and North Wales the variance (C_0) is considerably lower after a reduction by OSU86E. The converse is true for Derbyshire, however for nationwide consistency, OSU86E was chosen for use.

Table 1 Characteristics of the Reduced Data

	East Anglia		Derbyshire		North Wales	
	IFE	OSU	IFE	OSU	IFE	OSU
reduced mean	-1.7	-1.5	-4.8	-1.1	-6.9	-2.4
reduced st. dev.	6.6	4.6	11.3	13.0	17.8	17.8
(C_0) mgal ^{2}	46.3	23.9	150.5	171.6	358.1	318.0
Correlation Dist ξ^o	0.24	0.16	0.12	0.14	0.19	0.12
mean (obs)	-0.70		16.30		32.82	
st. dev (obs)	4.36		14.45		22.95	
npts	600		647		557	

When predicting heterogeneous quantities, such as geoid heights from gravity anomalies, it is imperative that the appropriate covariance function is used, to minimize prediction errors. An analytical model is fitted to the gravity anomaly empirical covariance function using an iterative adjustment of the model parameters (Knudsen, 1987). The covariance function however involves values for the error per degree of the geopotential model. These may be computed empirically but they are not available for the 360 sets, so two parametric models have been devised such that the error for degree 100 is 1 mgal (Tscherning, 1989).

Table 2 shows the model function parameters for the Derbyshire data after reduction by both reference models, and using different sets of error degree variances. It is obvious that the 360 model degrees must be accounted for even though models 2 and 3 are empirically derived. An assessment must now be made as to which model is best as the values do differ - the effect of this on the final predictions must be determined.

Table 2 Comparison of Error Degree Variance Models used in Fitting a Model Covariance Function in the Derbyshire Area

Parameter	IFE	OSU	IFE	OSU	IFE	OSU
Nmax	180	180	360	360	360	360
model	3	2	4	4	5	5
C_o	148.9	171.1	148.5	165.7	149.4	166.0
a	0.03	-0.4	0.03	0.01	0.02	0.06
$R-R_B$	+1.512	+2.894	+3.386	+7.348	+2.983	+7.092

where model 2 = Rapp's 1981 error degree variances
3 = GPM2 error degree variances
4 = linear parametric model
5 = quadratic parametric model

Topographic Effects

Much variation in the gravity field is due to the topography. It is anticipated that the contribution from the terrain will be quite significant in North Wales where the variation is over 300 mgal and the topography is high and rugged. At the moment the covariance function is irregular and the associated geoid prediction errors are not good. Conversely, the East Anglia area is flat and homogeneous, and the variation of the gravity field is small. Considering the topography here would have little effect if any.

Results

The results shown are from a solution for East Anglia using the geopotential model OSU86E, and free-air anomalies only.

Fig 4 The topography in North Wales, Contour Interval = 100 m

Table 3 A Comparison of 'Observed' with Predicted Geoid Height Differences, from the East Anglia Scheme

Stns	Dist	ΔH	Δh	ΔN^o	ΔN^p	$\varepsilon_{\Delta N}$
18-19	33.0	84.495	84.776	-0.281	-0.287	0.006
18-20	47.0	22.861	23.401	-0.540	-0.578	0.038
18-21	48.1	35.829	36.388	-0.559	-0.587	0.028
18-24	44.2	35.319	35.841	-0.522	-0.548	0.025
19-20	45.4	-61.683	-61.375	-0.2308	-0.291	-0.017
19-21	30.8	-48.861	-48.388	-0.293	-0.300	0.007
19-24	37.2	-49.211	-48.935	-0.276	-0.261	-0.015
20-21	22.2	12.969	12.987	-0.018	-0.009	-0.009
20-23	23.4	-0.632	-0.812	0.180	0.235	-0.055
20-24	9.2	12.465	12.440	0.025	0.030	-0.005
21-23	18.4	-13.595	-13.799	0.204	0.244	-0.040
21-24	7.4	-0.523	-0.547	0.024	0.039	-0.015
23-24	13.4	13.095	13.253	0.158	0.205	-0.047

ΔH = Ellipsoidal height differences (m)

Δh = Orthometric height differences (m)

ΔN^o = 'Observed' geoid height differences = DH - Dh (m)

ΔN^p = Predicted height differences (m)

The predicted geoid height differences were derived using covariance function parameters $C_o = 23.91\ mgal^2$ and $R - R_B = 7.5$ km. The GPS values were produced by an ambiguity fixed 'double difference' solution. It can be seen that the errors in the relative geoid height range from less than a centimetre to about 6 cm over distances of up to 50 km. These are expected to improve with further processing of the GPS data and a refinement of the geoid prediction method.

CONCLUSIONS

It can be concluded that a geoid database can be computed for Great Britain which will allow the use of GPS for levelling. The quality may decline in the highland areas even after a serious consideration of the terrain contribution. It is questionable however, if the need for such accuracy will arise in these areas if the database is for use in Engineering and topography surveys. A further problem is the island and coastal areas where predictions may suffer from not having a constant data distribution on all sides. The BGS is intending to undertake near-shore gravity surveys which would overcome this somewhat.

Acknowledgements

Research on precise relative geoid determination at Nottingham University is sponsored by the Science and Engineering Research Council, the Ordnance Survey and the British Geological Survey. Acknowledgements are also due to Professor C C Tscherning of the Kort-og Matrikelstyrelsen, Denmark.

REFERENCES

Arabelos, D (1989) 'Gravity Field Approximation in the Area of Greece with Emphasis on Local Characteristics'. Bulletin Geodesique, Vol 63, No 1 pp 69-84.

Benciolini, B, Mussio, L, Sanso, F, Gaspenni, P, Zerbini, S (1983) 'GEoid Computations in the Italian Area.

Knudsen, P (1987) 'Estimation and modelling of the Local Covariance Function using Gravity and Satellite Altimeter Data' Bulletin Geodesique Vol 61, pp 145-160.

Tscherning, CC 'Achievable Accuracy for Geoid Height Differences Across the Strait of Gibraltar'. First International Workshop on Geodesy for the Europe-Africa fixed Link Feasibility Studies in the Strait Gibralta, Madrid, 8-10 March 1989.

TOWARDS A CROSS–VALIDATED SPHERICAL SPLINE GEOID FOR THE SOUTH–WESTERN CAPE, SOUTH AFRICA

Herman van Gysen
Department of Surveying and Mapping
University of Natal, Durban, South Africa

Charles L Merry
Department of Surveying
University of Cape Town, Cape Town, South Africa

INTRODUCTION

Spline functions have interesting theoretical and practical properties that make them most useful and effective in a great many approximation problems, not least in geodetic problems such as local modelling of the gravity field. The application we consider here is local modelling of the geoid at the hand of given geoid heights (or height anomalies), and given free–air gravity anomalies. This is a problem of the type likely to be encountered if GPS is to be used for levelling, to obtain differences in orthometric height (or differences in normal height) where geoid heights are obtained from GPS measurements at points of known elevation. The gravity anomalies are used to help with the interpolation of the geoid between the GPS–derived geoid heights.

The appropriate spline functions to use are the spherical splines and harmonic splines whose theory is so elegantly developed by Freeden (1981a,b). From the wide family of spherical splines we will choose the spline function minimising the semi–norm

$$\|u\|^2_{R\Omega,2} = \int_{R\Omega} |\nabla^2_* u|^2 d\omega \tag{1}$$

among distributions on the sphere $R\Omega$ for which the surface Laplacian ∇^2_* is square integrable. The nullspace of the semi–norm is made up of constant functions. For this choice the spline 'degree variances' fall off eventually as n^{-3}, at the same rate as the degree variances of the disturbing potential in Tscherning and Rapp's widely used model (n is spherical harmonic degree). The corresponding kernel function needs to be regularised (by a Bjerhammar regularisation 'burying' the knots on an internal sphere of radius R_B) if gravity anomalies are to be admitted as data. The kernel function we shall use when regularised in this way is:

$$K_2(\xi\cdot\eta) = \sum_{n=1}^{\infty} \frac{2n+1}{4\pi}\{n(n+1)\}^{-2}\left(\frac{R_B}{R}\right)^{2n+2} P_n(\xi\cdot\eta), \tag{2}$$

where ξ,η are points on the unit sphere Ω, and R is the radius of a spherical earth.

The technique of generalised cross–validation (Golub *et al.*, 1979; Wahba and Wendelberger, 1980) provides an objective means of smoothing the data contributing to the spline representation, by minimising the prediction error at points removed one at a time from the data. At the same time the cross–validation provides access to regression diagnostics based on the same procedure of deleting single measurements in turn from the overall data set.

We will illustrate the concept of optimal spherical spline smoothing and the working of deletion diagnostics at the hand of a GPS survey of the University of Cape Town/Council for Scientific and Industrial Research (UCT/CSIR) Tygerberg Test Area in the south–western Cape, near Cape Town. The results of this GPS survey (a joint venture between UCT and the Institut für Angewandte Geodäsie, Frankfurt) are not yet available; we have simulated its results by taking geoid heights from a local gravimetric geoid.

SMOOTHING SPLINE GEOID

The kernel function (2), with the depth of the Bjerhammar sphere chosen according to data spacing by the rule suggested by Lelgemann (1981), acts in the nature of a low–pass filter removing from the spline representation the higher frequency component that cannot be recovered from the data. The data should similarly be smoothed, also in view of measurement error. We do this by looking for a spline that simultaneously minimises the semi–norm (1) and gives a least squares fit to the data.

Suppose geoid heights N_i are given at (distinct) data points $\alpha_i = R\nu_i$, $\nu_i \epsilon \Omega$, $i=1,\dots,N$, on a spherical earth of radius R; and gravity anomalies Δg_j are given at (distinct) points $\beta_j = R\mu_j$, $\mu_j \epsilon \Omega$, $j=1,\dots,M$. The required spline has the form

$$s(x) = \sum_{i=1}^{N} a_i K_{2,B}(\xi\cdot\nu_i) + \frac{1}{\gamma R_B} \sum_{j=1}^{M} b_j \Delta_g K_{2,B}(\xi\cdot\mu_j) + c_0, \tag{3}$$

where $x=R\xi$ is a point on the sphere $R\Omega$, c_0 is a constant (spherical harmonics of degree zero are in the nullspace of the semi–norm), and where

$$\Delta_g k_n = (n-1)k_n, \; n=1,2,\dots, \tag{4}$$

relates (except for the otherwise usual factor R) the 'degree variances' for the representation of the disturbing potential to the degree 'cross–covariances' between disturbing potential and gravity anomalies. γ is an average value of gravity.

Let $\lambda>0$; the required spline minimises

$$\lambda \int_{R\Omega} |\nabla_*^2 u|^2 d\omega + \mathbf{r}^T \mathbf{Q}^{-1} \mathbf{r} \tag{5}$$

among distributions whose surface Laplacians are square integrable, where $\mathbf{Q}$ is the 'cofactor matrix' differing from the measurement variance matrix by a factor σ^2. The analysis of the next section requires $\mathbf{Q}$ to be a diagonal matrix, but σ need not be known. The vector $\mathbf{r}$ contains residuals, $r_i = u(\alpha_i) - N_i$, $i=1,\ldots,N$, for geoid height measurements, and $r_{N+j} = (1/\gamma R)\Delta_g u(\beta_j) - \Delta g_j$, $j=1,\ldots,M$, for gravity anomalies.

The factor λ directs the amount of trade–off between smoothing and fidelity to the data. In the extreme cases, $\lambda=0$ directs the solution to strict interpolation, and $\lambda=\infty$ requires the solution to come from the spline nullspace, i.e. the solution is a constant function.

The coefficients of the representation (3) are found from the data by solving the linear equations:

$$\sum_{i=1}^{N} a_i K_{2,B}(\nu_k \cdot \nu_i) + \frac{1}{\gamma R_B} \sum_{j=1}^{M} b_j \Delta_g K_{2,B}(\nu_k \cdot \mu_j) + \lambda q_{kk} + c_0 = N_k, \quad k=1,\ldots,N;$$

$$\frac{1}{\gamma R} \sum_{i=1}^{N} a_i \Delta_g K_{2,B}(\mu_l \cdot \nu_i) + \frac{1}{\gamma^2 R R_B} \sum_{j=1}^{M} b_j \Delta_g^2 K_{2,B}(\mu_l \cdot \mu_j) + \lambda q_{N+l,N+l} - \frac{c_0}{\gamma R} = \Delta g_l,$$

$$l=1,\ldots,M; \quad (6)$$

subject to the 'natural spline end condition':

$$\sum_{i=1}^{N} a_i - \frac{1}{\gamma R} \sum_{j=1}^{M} b_j = 0. \quad (7)$$

These equations can be written more compactly as:

$$\begin{aligned} \mathbf{Ma} + \mathbf{B}^{\mathrm{T}}\mathbf{b} &= \mathbf{c}; \\ \mathbf{Ba} &= \mathbf{0}, \end{aligned} \quad (8)$$

where $\mathbf{M} = \mathbf{A} + \lambda \mathbf{Q}$, and

$$\mathbf{A} = \begin{bmatrix} K_{2,B}(\nu_k \cdot \nu_i) & \frac{1}{\gamma R_B} \Delta_g K_{2,B}(\nu_k \cdot \mu_j) \\ \frac{1}{\gamma R} \Delta_g K_{2,B}(\mu_l \cdot \nu_i) & \frac{1}{\gamma^2 R R_B} \Delta_g^2 K_{2,B}(\mu_l \cdot \mu_j) \end{bmatrix}$$

for rows $k=1,\ldots,N$; $l=1,\ldots,M$; and columns $i=1,\ldots,N$; $j=1,\ldots,M$;

$$\mathbf{B} = (1,\ldots,1; -\frac{1}{\gamma R},\ldots, -\frac{1}{\gamma R}),$$

i.e., N ones followed by M times $-(1/\gamma R)$;

$$\mathbf{a}^{\mathrm{T}} = (a_1,\ldots,a_N; b_1,\ldots,b_M);$$
$$\mathbf{b} = c_0;$$

$$\mathbf{c}^{\mathrm{T}} = (N_1,\dots,N_N; \Delta g_1,\dots,\Delta g_M).$$

The vectors **a** and **b** contain the coefficients of the spline fit and **c** contains the measurements.

The solution of (8) is:

$$\mathbf{a} = \mathbf{M}^{-1}\mathbf{c} - \mathbf{M}^{-1}\mathbf{B}^{\mathrm{T}}\mathbf{b};$$
$$\mathbf{b} = (\mathbf{B}\mathbf{M}^{-1}\mathbf{B}^{\mathrm{T}})^{-1}\mathbf{B}\mathbf{M}^{-1}\mathbf{c}. \tag{9}$$

The existence of the solution is guaranteed. The values the spline takes at the data points may be calculated from $\mathbf{s} = \mathbf{Aa} + \mathbf{B}^{\mathrm{T}}\mathbf{b}$, or from $\mathbf{s} = \mathbf{Hc}$, where

$$\mathbf{H} = \mathbf{A}\mathbf{M}^{-1}(\mathbf{I} - \mathbf{B}^{\mathrm{T}}(\mathbf{B}\mathbf{M}^{-1}\mathbf{B}^{\mathrm{T}})^{-1}\mathbf{B}\mathbf{M}^{-1}) + \mathbf{B}^{\mathrm{T}}(\mathbf{B}\mathbf{M}^{-1}\mathbf{B}^{\mathrm{T}})^{-1}\mathbf{B}\mathbf{M}^{-1}. \tag{10}$$

Here **H** is the 'hat matrix' for smoothing splines; it has many of the properties of the hat matrix of linear least squares, but is not a projection matrix (Eubank, 1984). There is a more attractive formula for calculating **H** when the spline representation is given in terms of reproducing kernel functions (Wahba, 1978).

The smoothing parameter λ is chosen to minimise mean square prediction error. An optimal value can be obtained from the data by removing points one at a time and choosing the value of λ that best predicts the data values at the missing points. The score to be minimised is thus

$$F(\lambda) = \sum_{i=1}^{N} \frac{(s^{-i}(\alpha_i)-N_i)^2}{\sigma^2_{ii}} + \sum_{j=1}^{M} \frac{(\frac{1}{\gamma R}\Delta_g s^{-j}(\beta_j)-\Delta g_j)^2}{\sigma^2_{N+j,N+j}}, \tag{11}$$

where $s^{-i}(\alpha_i)$ is the value at α_i of the smoothing spline calculated from all the data except the observation at the point α_i, and similarly for the gravity anomaly $\frac{1}{\gamma R}\Delta_g s^{-j}(\beta_j)$. It is not difficult to show that

$$s^{-i}(\alpha_i) - N_i = \frac{s(\alpha_i)-N_i}{1-h_{ii}},\ i=1,\dots,N; \tag{12}$$

$$\frac{1}{\gamma R}\Delta_g s^{-j}(\beta_j) - \Delta g_j = \frac{\frac{1}{\gamma R}\Delta_g s(\beta_j)-\Delta g_j}{1-h_{N+j,N+j}},\ j=1,\dots,M; \tag{13}$$

where h_{ii} are elements of the hat matrix **H**. The derivation follows an argument familiar from sequential least squares estimation (see Robinson and Moyeed, 1989).

Thus (11) could equally well be written as

$$F(\lambda) = \sum_{i=1}^{N} \frac{(s(\alpha_i)-N_i)^2}{\sigma_{ii}^2(1-h_{ii})^2} + \sum_{j=1}^{M} \frac{(\frac{1}{\gamma R}\Delta_g s(\beta_j)-\Delta g_j)^2}{\sigma_{N+j,N+j}^2(1-h_{N+j,N+j})^2} = \sum_{i=1}^{n} \frac{r_i^2}{\sigma_{ii}^2(1-h_{ii})^2}, \tag{14}$$

where the residuals come from $\mathbf{r}=(\mathbf{I}-\mathbf{H})\mathbf{c}$, and $n=N+M$. In situations where σ is not known Craven and Wahba (1979) and Golub *et al.* (1979) show that the generalised cross–validation score is appropriate on both theoretical and practical grounds:

$$V(\lambda) = \sum_{i=1}^{n} \frac{r_i^2}{q_{ii}}(\frac{1}{n}\mathrm{trace}(\mathbf{I}-\mathbf{H}))^{-2}. \tag{15}$$

An example of the use of (15) in the spherical spline context is given by Wahba and Wendelberger (1980). This is the function we seek to minimise.

DIAGNOSTICS

As with problems in linear least squares estimation, analysis of the spline smoothing problem can be extended to include considerations of reliability and the possibility of outliers among the data. The key to this analysis is the influence matrix **H**. Through it observations can be identified that are especially influential, or exert especial leverage, forcing the data to an interpolatory fit.

As measure of influence we use $h_{ii}r_i/(1-h_{ii})$, which gives the change in fit at the ith point upon deleting the measurement at that point. This deleted from fit (*DFFIT*) criterion comes from the diagnostic analysis of linear regression problems (Belsley *et al.*, 1980, p. 15). When scaled by dividing by $\sigma\sqrt{h_{ii}}$, the standard error of the fit, the scaled deleted from fit diagnostic measure is obtained:

$$DFFITS(i) = (\frac{h_{ii}}{1-h_{ii}})^{1/2} \frac{r_i}{\hat{\sigma}(1-h_{ii})^{1/2}}, \tag{16}$$

where the variance is estimated through

$$\hat{\sigma}^2 = \sum_{i=1}^{n} \frac{\mathbf{r}^T\mathbf{Q}^{-1}\mathbf{r}}{\mathrm{trace}(\mathbf{I}-\mathbf{H})} \tag{17}$$

(Silverman, 1985; Eubank, 1985), or since the observation under review contributes to the variance, Eubank (1985) shows how it may also be calculated from deleted residuals. The denominator in (17) may be regarded as the 'equivalent degrees of freedom for smoothing splines' (Wahba, 1983; Eubank, 1985), so one may look for residuals exceeding a Student's t value with trace(**I**–**H**) degrees of freedom.

A graph of influence $x=[h_{ii}/(1-h_{ii})]^{1/2}$ versus standardised residual $z=r_i/\hat{\sigma}(1-h_{ii})^{1/2}$ permits identification of points with high leverage and (typically) small residuals, or low leverage points with (possibly insignificant) large residuals.

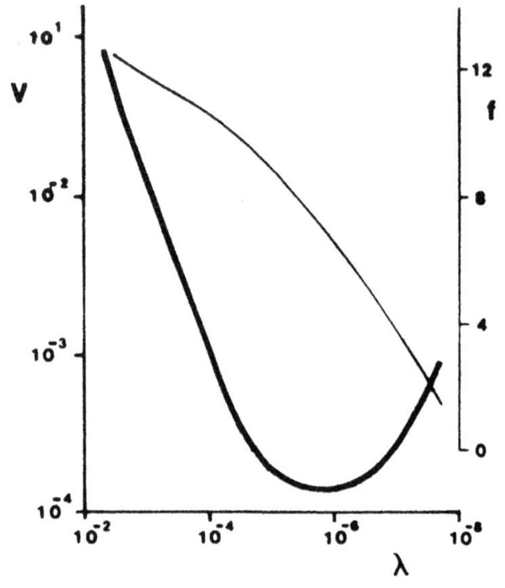

Fig. 1. Graph of cross–validation score $V(\lambda)$ versus smoothing parameter λ for the smoothing spline fit to simulated geoid heights at the 14 points of the UCT/IfAG GPS survey. The auxiliary scale shows effective degrees of freedom (out of a maximum possible 13).

Belsley *et al.* (1980) give some advice on cut–off values for $DFFITS(i)$, and suggest $2\sqrt{(n-f)/n}$, where f is the number of degrees of freedom.

TOWARDS A SPLINE GEOID FOR THE WESTERN CAPE

An optimal smoothing factor and diagnostics are sought for the smoothing spline geoid for a part of the south–western Cape Province, in the area of the UCT/CSIR Tygerberg Test Area established by Professor H.S. Williams of the University of Cape Town. This is a high–precision three–dimensional geodetic network which has also been surveyed by GPS, in November 1988, in a joint venture between the Institut für Angewandte Geodäsie, Frankfurt, and the University of Cape Town. Results for the GPS survey are not yet available; here we simulate the results by taking geoid heights from a local geoid for the same area (Van Gysen, 1986). Gravity data are taken from (Merry, 1983).

In the calculations that follow the Bjerhammar regularising factor will be set to $R_B/\mathrm{R}=0{,}9994$.

Spline geoid from geoid heights only. We first attempt an optimally smoothed spline geoid for the test area on the basis of simulated geoid heights at the 14 stations of the UCT/IfAG GPS survey. These heights were given (somewhat arbitrarily) an *a priori* standard error of 3 cm. Figure 1 shows the graph of predicted error V of the cross–validatory spline fit for different choices of the smoothing parameter λ. Also shown in the figure are the effective degrees of freedom for each fit – these range from a maximum of 13 for large values of λ when the spline fit reduces to the arithmetic mean of the data), to a value of zero for an interpolating fit. The optimal value of V gives a fit with 7,85 degrees of freedom, showing that there is a certain amount of redundancy in the data.

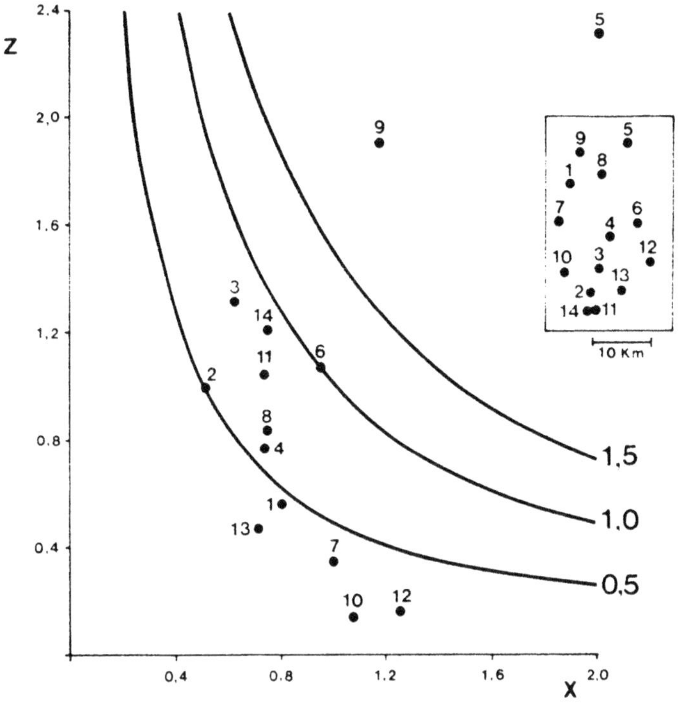

Fig. 2. *DFFITS* diagnostic plot for the smoothing spline fit to the 14 geoid simulated geoid heights of the UCT/IfAG GPS survey. $(x,z) = \{[h_{ii}/(1-h_{ii})]^{1/2}, r_i/s(1-h_{ii})^{1/2}\}$. The inset shows the relative location of the 14 points.

For a point by point appraisal of the fit we use the *DFFITS* diagnostic shown in Figure 2. The x–axis gives $[h_{ii}/(1-h_{ii})]^{1/2}$ and the z–axis the standardised residuals $r_i/\hat{\sigma}(1-h_{ii})^{1/2}$. Point 5 'on its own' near the top right corner of the network is clearly a leverage point, as is point 9. The *DFFITS* cut–off for the network is 1,3, on which basis the data values at these points warrant inspection (despite the fact that the raw residuals appear innocuous).

Spline geoid from geoid heights and gravity anomalies. Six of the 14 geoid heights above were taken through into the next calculation (gravity anomalies contribute relatively more when there are fewer values of N), and 24 values of Δg randomly selected from the 100 or so in the area of the GPS survey. The geoid heights were given an *a priori* standard error of 3 cm, and the free–air anomalies an error of 0,5 milligal. The graphs corresponding to Figures 1 and 2 are not shown here; that of of V versus λ has much the same shape as in Figure 1. The optimal fit has 7,15 degrees of freedom, indicating a model somewhat sensitive to the data. As is to be expected, the geoid heights are influential, especially at points at the outer edges. Gravity anomalies contribute less to the geoid model. Nevertheless, it is our experience that in situations where few geoid heights are given, and provided the trouble is taken to choose an appropriate smoothing value, supplementary gravity data can contribute significantly to an improved local representation of the geoid (Van Gysen, 1988).

Spherical splines do work for local modelling of the geoid, almost as well as least squares collocation. The cross validation technique provides an objective means of choosing an optimal degree of smoothing, and provides access to diagnostics for reliability and outliers.

Acknowledgement. We are grateful to Professor H.S. Williams for providing data on the UCT/CSIR Tygerberg Test Area and details of the UCT/IfAG GPS survey.

REFERENCES

Belsley, D.A., Kuh, E. and Welsch, E. (1980). *Regression diagnostics: identifying influential data and sources of collinearity*, Wiley, New York.

Craven, P. and Wahba, G. (1979). Smoothing noisy data with spline functions, *Numer. Math.*, **31**, 377–403.

Eubank, R.L. (1984). The hat matrix for smoothing splines, *Statist. and Probab. Letters*, **2**, 9–14.

Eubank, R.L. (1985). Diagnostics for smoothing splines, *J. Royal Statist. Soc. B*, **47(2)**, 332–341.

Freeden, W. (1981a). On approximation by harmonic splines, *Manus. Geod.*, **6(2)**, 193–244.

Freeden, W. (1981b). On spherical spline interpolation and approximation, *Math. Meth. Appl. Sci.*, **3**, 551–575.

Golub, G.H., Heath, M. and Wahba, G. (1979). Generalized cross–validation as a method for choosing a good ridge parameter, *Technometrics*, **21(2)**, 215–223.

Lelgemann, D. (1981). On numerical properties of interpolation with harmonic kernel functions, *Manus. Geod.*, **5**, 157–191.

Merry, C.L. (1983). Gravity survey of degree square 3318, University of Cape Town, Dept. of Surveying, Internal Report G–10.

Robinson, T. and Moyeed, R. (1989). Making robust the cross–validatory choice of smoothing parameter in spline smoothing regression, *Comm. Statist. – Theor. Meth.*, **18(2)**, 523–589.

Silverman, B.W. (1985). Some aspects of the spline smoothing approach to non–parametric regression curve fitting (with discussion), *J. Roy. Statist. Soc. B*, **47(1)**, 1–52.

Van Gysen, H. (1986). A relatively precise geoid for the western Cape: preparations for GPS levelling, *Proc. Symposium on Geodetic Positioning for the Surveyor*, Cape Town, 143–152.

Van Gysen, H. (1988). Splines and local approximation of the earth's gravity field, University of Cape Town, unpublished PhD thesis.

Wahba, G. (1978). Improper priors, spline smoothing and the problem of guarding against model errors in regression, *J. Roy. Statist. Soc. B*, **40(3)**, 364–372.

Wahba, G. (1983). Bayesian 'confidence intervals' for the cross–validated smoothing spline, *J. Roy. Statist. Soc. B*, **45(1)**, 364–372.

Wahba, G. and Wendelberger, J. (1980). Some new mathematical methods for variational objective analysis using splines and cross–validation, *Monthly Weather Review*, **108**, 1122–1143.

A GRAVIMETRIC GEOID IN TAIWAN AREA

Ruey-Gang Chang, Chia-Chyang Chang and Jenn-Taur Lee
Dept. of Surveying and Mapping Engineering
Chung Cheng Institute of Technology, Tashi, Taoyua (33509)
Taiwan

ABSTRACT

The paper describes the theoretical aspects of the gravimetric geoid. Stokes' integral convolution approach reformulated for the higher-order reference field, OSU86F in this case, is used. In addition to the theoretical aspects, an example of the actual solution for Taiwan area is shown.

INTRODUCTION

The geoid can be determined either directly from observations or derived from other gravity field parameters (from terrestrial gravity anomalies Δg, from satellite derived potential coefficients, and from astrogeodetic deflections of the vertical ξ, η). However these determinations are not equivalent. Some geodesists suggest that combinations of the various kinds of data should be used. By properly combining different kinds of data, and exploiting the better global homogeneity inherent in one data set on the one hand; one should be able to enhance the quality of the result. Clearly, there are numberous ways to properly combine heterogeneous data to advantage.

The computation of the geoid to an accuracy of 10 cm become a practical requirement in the present. Several studies have been carried out to incorporate different combinations of data sets, e.g. Vanicek et al. (1987).

The purpose of this paper is to discuss some of the procedures used for the gravimetric geoid in Taiwan area. This paper is also to consider the systematic effects and improve the accuracy in the computation.

Our reference ellipsoid is the Geodetic Reference System 1967 (GRS67). For saving computation time, we are only interested in the area bounded by latitude and longitude 21.5° N to 25.5° N, 120° E to 122° E.

IMPROVEMENTS ON THE GEOIDAL UNDULATION

The Atmospheric Correction

Anomaly Correction for the Atmosphere. In the case of Stokes' equation the boundary value problem of physical geodesy is based on the assumption of no masses outside the boundary surface, i.e., the geoid, for only then is Laplace's equation to zero.

In order to consider the situation, Moritz (1974) proposed a technique where the atmosphere is redistributed inside the reference in a symmetric way. This redistribution will cause changes δg_A in gravity and δW_A in gravity potential. Then,

$$g = g^\circ + \delta g_A \,, \qquad (1)$$

$$W = W^\circ + \delta W_A \,, \qquad (2)$$

where g and W are the gravity and geopotential in the case of points on the real earth with atmosphere, while g° and W° are value referred to the system where the atmosphere is redistribnuted. In this procedure a gravity anomaly and geoid undulation would be

$$\Delta g = \Delta g^\circ + \delta g_A\,, \qquad (3)$$

$$N = N^\circ + \delta W_A/g\ , \qquad (4)$$

It can be shown that $\delta W_A/g$ for points on the geoid is approximately 6 mm and thus negligible for the computations. These values are always positive, values of $-\delta g_A$ have been computed by Ecker and Mittermayer (1969) and are also tabulated in the work Geodetic Reference System, 1967 (IAG, p.72). This treatment implies that the effect on observed gravity g of all the atmosphere at elevation above that of the station at the surface of the Earth, could be evaluated as being due to series of homogeneous ellipsoid shells.

Atmospheric effect in a combination data. In order to achieve a consistent system of data for the estimation of potential coefficients from satellite observations and terrestrial gravity anomalies the latter have to be corrected for the atmosphere. Thus in the Earth model determinations the atmospheric term δg_A should be applies to all gravity anomalies to obtain Δg°, which is the anomaly to use in combination solutions.

From a spherical harmonic development of δg_A to degree and order 360 (i.e. OSU86F) we obtained corrections $\bar{C}_{nm}{}^A$ and $\bar{S}_{nm}{}^A$:

$$\bar{C}_{nm}{}^A = -\,(n-1)\ \delta g_{A,nm}{}^{\bar{C}}/\gamma\,,$$

$$\bar{S}_{nm}{}^A = -\,(n-1)\ \delta g_{A,nm}{}^{\bar{S}}/\gamma\,, \qquad (5)$$

where

$$\delta g_{A,nm}{}^{\bar{C}} = (\textstyle\iint_\sigma \delta g_A \bar{P}_{nm}(\sin\bar{\phi})\cos m\lambda d\sigma)/4\pi\ ,$$

$$\delta g_{A,nm}{}^{\bar{S}} = (\textstyle\iint_\sigma \delta g_A \bar{P}_{nm}(\sin\bar{\phi})\sin m\lambda d\sigma)/4\pi\ , \qquad (6)$$

$\bar{P}_{nm}(\sin\bar{\phi})$ is associated Legendre functions, as a funstion of geocentric latitude $\bar{\phi}$,

γ is mean value of the gravity over the earth,

$\bar{\phi}$ is geocentric latitude of the computation point,

λ is geodetic longitude of the computation point.

The coefficients $\bar{C}_{nm}^A$ and $\bar{S}_{nm}^A$ would have to be added to the coefficients obtained from terrestrial gravity anomalies not corrected for the atmosphere. Coefficients $\bar{C}_{nm}^A$ and $\bar{S}_{nm}^A$ of magnitude 1×10^{-10} occur up to degree 19 (Rapp and Rummel, 1975).

The zero-order term of eqn.(6):

$$\delta g_{A,00}{}^{\bar{C}} = (\textstyle\iint_\sigma \delta g_A d\sigma)/4\pi\ , \qquad (7)$$

represents the $M(\delta g_A) = \mu_c$ of δg_A over the earth and is (Vanicek and Krakiwsky, 1982) :

$$\delta g_{A,00}{}^{\bar{C}} = \mu_c = -0.853\ \text{mGal}.$$

Atmospheric Correction in Geoid Computations. If the geoid computation is basically owing to (Molodenskij et al, 1962) :

$$N = R(\iint_{\sigma_c} S(\psi)\Delta g d\sigma_c)/4\pi\gamma + R(\sum_{n=2}^{\infty} Q_n(\psi_c)\Delta g_n(\bar{\phi},\lambda))/2\gamma, \tag{8}$$

where

$Q_n(\psi_c)$ is the truncation coefficients for ψ_c,

R is mean radius of the earth ,

$S(\psi)$ is Stokes' function for the spherical distance ,

and

$$\Delta g_n(\bar{\phi},\lambda) = \gamma(n-1)\sum_{m=0}^{n}(\bar{C}_{nm}^* \cos m\lambda + \bar{S}_{nm}^* \sin m\lambda)\bar{P}_{nm}(\sin \bar{\phi}) . \tag{9}$$

Since gravity anomalies are not available over the entire earth, geoid computations are carried out by combing terrestrial gravity anomalies given in a cap σ_o of spherical radius ψ_c with potential coefficients derived from satellite observations. The gravity anomaly can be defined:

$$\Delta g = g - \gamma_A , \tag{10}$$

where γ_A is normal gravity which derived from geodetic constants that include the mass of earth and atmosphere.

The gravity anomaly corrected for atmosphere is, from eqn.(3)

$$\Delta g^\circ = g - \delta g_A = g - \delta g_A - \gamma_A , \tag{11}$$

We define the mean value of Δg° over the earth, which is given in (Rapp and Rummel, 1975):

$$\Delta g_o = M(\Delta g^\circ) = (\iint_\sigma \Delta g^\circ d\sigma)/4\pi , \tag{12}$$

Ideal gravity anomalies Δg_I are defined to be those which their global mean is zero. Such a value can be obtained by removing Δg_o from Δg° . We have

$$\Delta g_I = \Delta g^\circ - \Delta g_o = \Delta g - \delta g_A - \Delta g_o , \tag{13}$$

Then eqn. (8) becomes to

$$N = R(\iint_{\sigma_c} S(\psi)\Delta g_I d\sigma_c)/4\pi\gamma + R(\sum_{n=2}^{\infty} Q_n(\psi_c)\Delta g_n(\bar{\phi},\lambda))/2\gamma, \tag{14}$$

Inserting eqn.(13) to (14), we find

$$N = R(\iint_{\sigma_c} S(\psi)\Delta g d\sigma_c)/4\pi\gamma + R(\sum_{n=2}^{m} Q_n(\psi_c)\Delta g_n(\bar{\phi},\lambda))/2\gamma$$
$$- R(\iint_{\sigma_c} S(\psi)\, \delta g_A\, d\sigma_c)/4\pi\gamma - R(\iint_{\sigma_c} S(\psi)\, \Delta g_o\, d\sigma_c)/4\pi\gamma , \tag{15}$$

Following Lambert and Daring (1936):

$$\Phi = (\iint_{\sigma_c} S(\psi)d\sigma_c)/4\pi = (\int_{\sigma_c} S(\psi)\sin\psi d\psi)/2 , \tag{16}$$

We finally obtain:

$$N = R(\iint_{\sigma_c} S(\psi)\Delta g d\sigma_c)/4\pi\gamma + R(\sum_{n=2}^{m} Q_n(\psi_c)\Delta g_n(\phi,\lambda))/2\gamma - R(\mu_c\Phi(\psi_c))/\gamma . \tag{17}$$

where μ_c is a constant for δg_A within the cap.

To compare (17) with (8), then we can get atmospheric correction term N_A :

$$N_A = - R(\mu_c\Phi(\psi_c))/\gamma . \tag{18}$$

To estimate the effect of the atmosphere on the geoid computations, we consider several cap sizes in caps having three different mean elevations of 0 m, 100 m, and 200 m, where values of δg_A are -0.87 mGal, -0.85 mGal, and -0.85 mGal, respectively.

The Topographic Effect

Our observations on the surface of the earth are affectecd very strongly by topography; this effect increases with increasing distance from the geoid, i.e., as altitude increases. In fact, compared with a model vertical gradient, the actual vertical gradient of gravity within the earth is always larger in absolute value in the presence of topographical relief than when the surface is flat. This phenomenon is most expediently evaluated through the topographical effect δg_T on observed gravity.

Having established that squashing the Bouguer plate on to the geoid makes no difference on the observed gravity, we can squash the outside masses (outside the geoid) onto the geoid and assumed as if there were no masses outside the geoid. This results in a situation where the observed gravity g is reduced to the geoid by means of free air reduction. The superior-position of terrain on the Bouguer plate will result in the change of Bouguer plate attraction by δg^{BT}. On the other hand, the superior-position of density irregularity on the surface layer on the geoid results in the effect of δg^{LT}. Hence the total effect of having squashed this irregular topography should be $c = \delta g^{LT} - \delta g^{BT}$. Then the topographic correction to free air anomaly equals (Vanicek et al., 1987) :

$$\delta g_T = \delta g^{BT} - \delta g^{LT} ,$$
$$= G\rho R^2(\Sigma(\bar{H}_i^2 - H_A^2)\sigma_i/\bar{l}^3)/2 , \tag{19}$$

where

- G is the gravitational constant,
- R is mean earth radius,
- ρ is density,
- $\bar{l}_i$ is the mean distance of the ith 10' by 10' cell from the point of interest,
- $\bar{H}_i$ is the corresponding mean topographical height,
- H_A is the topographical height at the interested point P_A,
- σ_i is the surface element.

The Indirect Effect

One problem in computing the geoid from free air gravity anomalies concerns the setting of the Boundary Value Problem (BVP) of geodesy and the computation of the free air gravity anomalies. The BVP calls for no masses external to the geoid. The effect on the computed cogeoid of the unaccounted for masses between the geoid and the terrain is the indirect effect (δg_I). This effect has to be removed from the computed cogeoid.

The indirect effect can be removed mathematically if we know the shape $N(\phi, \lambda)$ of the geoid or the cogeoid with respect to the reference ellipsoid, and postulate an average density ρ of the unaccounted for masses (Vanicek and Krakiwsky, 1982). The indirect effect may be expressed in harmonic function so that through analysis it can be determined whether this effect is significant in the long or short wavelength components of the geoid, which is given by (Vanicek et al., 1987) :

$$N = \pi G\rho H_A^2/\gamma - (G\rho R^2(\Sigma(\bar{H}_i^3 - H_A^3)\sigma_i/\bar{l}^3))/6\gamma , \tag{20}$$

where $\bar{H}_i$, H_A and $\bar{l}i$ as in eqn.(19).

The Ellipsoidal Correction

The usual application of the Stokes' equation for geoid undulation computations using gravity anomalies is a spherical approximation (Heiskanen and Moritz, 1967). The error of the spherical approximation has been investigated by several authors who have sought a solution to the boundary value problem with the ellipsoid as a reference surface. For example, Molodenskij et al. (1962) describes a solution of the basic integral equation retaining equation terms of the order of the flattening. Lelgemann (1969) developed techniques to compute corrections to the undulations computed from the Stokes' equation.

He showed that the root mean square correction was ±20 cm with a maximum correction of about 60 cm. A somewhat revised approach to this approximation problem is discussed by Moritz (1980). Mather (1973) has also described a solution to the gravimetric boundary problem considering the ellipsoid.

Today, however, geoid undulations are not usually computed through global integration procedures. Instead, potential coefficient information is combined with gravity anomaly information in a cap surrounding the computating point. For this procedure we therefore cannot use directly the correction equations developed by Lelgemann. Instead, we must develop new equations.

We consider the ellipsoid form of the Stokes' equation as given by Moritz (1980, eqns.49-26) :

$$N_E = R(\iint_\sigma(\Delta g - e^2 \Delta g')S(\psi)d\sigma)/4\pi\gamma + e^2(1/4 + (3\sin^2\phi)/4)N , \qquad (21)$$

where

N is the approximate value of the geoid undulation,

$$\Delta g' = (\Sigma\Sigma(G_{nm} \cos m\lambda + H_{nm} \sin m\lambda)P_{nm}(\sin \phi))/R . \qquad (22)$$

Rapp (1981) indicated a significant improvement in ellipsoidal correction :

1. For the Stokes' equation :

$$\delta N_E = (e^2/2\gamma)(\sum_{n=0}^{\infty}(Q_n - X_n)\sum_{m=0}^{n}(G_{nm} \cos m\lambda + H_{nm} \sin m\lambda)P_{nm}(\sin \bar{\phi}))$$
$$+ e^2(1/4 + (3/4)\sin^2\phi)N , \qquad (23)$$

where

e is the first eccentricity,

Q_n is the Molodenskij coefficient,

N is the approximate value of the geoid undulation,

a is earth equatorial radius used in potential coefficient determination,

GM is geodetic gravitational constant,

C_{nm}, S_{nm} are potential coefficients;

$$X_n = \begin{cases} 0 & \text{if } n < 2 , \\ 2/(n-1) & \text{if } n \geq 2 , \end{cases}$$

$$\begin{Bmatrix} G_{nm} \\ H_{nm} \end{Bmatrix} = K_{nm} \begin{Bmatrix} A_{n-2,m} \\ B_{n-2,m} \end{Bmatrix} + \lambda_{nm} \begin{Bmatrix} A_{nm} \\ B_{nm} \end{Bmatrix} + \mu_{nm} \begin{Bmatrix} A_{n+2,m} \\ B_{n+2,m} \end{Bmatrix} ,$$

$$K_{nm} = (-3(n - 3)(n - m - 1)(n - m))/(2(2n - 3)(2n - 1)) ,$$

$$\lambda_{nm} = (n^3 - 3m^2n - 9n^2 - 6m^2 - 10n^2 + 9)/(3(2n + 3)(2n - 1)) ,$$

$$\mu_{nm} = (-(3n + 5)(n + m +2)(n + m + 1))/(2(2n + 5)(2n + 3)) ,$$

2. For the Modified Stokes' equation

If the Stokes function is modified by substracting $S(\cos\psi_o) = S_o$ from $S(\psi)$, then

$$\delta N_E = (e^2/2\gamma)(\sum_{n=0}^{\infty}(\bar{Q}_n - X_n)\sum_{m=0}^{n}(G_{nm}\cos m\lambda + H_{nm}\sin m\lambda)\, Pnm(\sin\bar{\phi}) + e^2(1/4 - (3/4)\sin^2\phi)N) \ , \qquad (24)$$

where $\bar{Q}_n = Q_n + (S_o/(n-1))(P_{n-1}(\cos\psi_o) - \cos\psi_o\, P_n(\cos\psi_o))$, $n \geq 1$,

We now will evaluate the ellipsoidal corrections terms in eqns.(23) and (24). Our start in potential coefficients are those of OSU86F (1986) taken to degree 360. This evaluation was done for several cap sizes. The maximum and root mean square correction are given in Table 1.

Table 1. Magnitude of ellipsoidal correction term.

ψ	0°	10°	20°	180°
rms / Method	Max rms (cm)	Max rms (cm)	Max rms (cm)	Max rms (cm)
Stokes	34 ± 10	-17 ± 12	-42 ± 10	-78 ± 10
Modified Stokes	34 ± 10	-14 ± 12	-42 ± 18	-78 ± 23

THE GRAVIMETRIC GEOID

The gravimetric geoid, represented by its undulations N_G with regard to a reference ellipsoid, is the geoid computed from observed gravity. The computational procedure can be visualized as a partial solution of the geodetic boundary value problem (e.g., (Moritz, 1980)). The principle difficult with these solution in practice has been the formal requirement of continuous gravity covering the entire geoid.

If gravity anomalies were available everywhere, both the medium and long wavelength components of the gravity field would be determinable. The lack of global gravity coverage has let to the approximation whereby the integration is limited to a sphere cap. The result of this is that only the short wavelength components of the gravity field can be determined by the terrestrial gravity anomalies within the sphere cap (Vanicek and Krakiwsky, 1982), while the long wavelength components might be determined by some other ways.

The Choice of Potential Field Model

Out of the variety of potential field models published so far, we have chosen to adopt the OSU86F coefficients set (Rapp and Cruz, 1986) as probably the most complete solution available. More recently published gravity field models, tailored for special applications in satellite geodesy, seem to be inadequate for our purposes.

We transfer the OSU86F spheroid undulations relative to the GRS67 ellipsoid. The differences of the model gravitational potential are:

$$J_{20} = -0.484165998 \times 10^{-6},$$
$$J_{40} = 0.541413790 \times 10^{-6},$$
$$J_{60} = -0.150012899 \times 10^{-6}.$$

Fig.1 shows the medium wavelength gravimetric geoid which produced using the OSU86F geopotential coefficients.

Application of the Molodenskij Procedure to the Stokes Kernel

The computation of geoidal height from Stokes' formula involves the integration of the gravity anomaly weighted by Stokes' function over the entire surface of the spherical earth. The formula for that computation is :

$$N = (R/4\pi\gamma)\iint S(\psi)\Delta g(\psi,\alpha)\sin\psi \, d\psi \, d\alpha \ , \qquad (25)$$

where R, γ, g, ψ as defined in eqns.(6) and (8).

Consider the eqn.(25) having the kernel function :

$$S(\psi) = \sum_{n=2}((2n+1)/(n-1))P_n(\cos\psi) \ , \qquad (26)$$

If the integration of eqn.(25) is extended only up to a certain spherical distance ψ_0 from the computation point P, then eqn.(25) can be written as :

$$N = (R/4\pi\gamma)\int_0^{2\pi}\int_0^{\psi_0} S(\psi)\Delta g(\psi,\alpha)d\sigma + \delta N_{TR} \ , \qquad (27)$$

where $d\sigma$ is the incremental surface element, and N_{TR} is the cap truncation error defined as

$$\delta N_{TR} = (R/4\pi\gamma)\int_0^{2\pi}\int_{\psi_0}^{\pi} S(\psi)\Delta g(\psi,\alpha)d\sigma \ , \qquad (28)$$

Following the practice given in (Heiskanen and Moritz, 1967), it defines the truncation function as

$$\bar{S}(\psi) = \begin{cases} 0 & \text{if } 0 \le \psi < \psi_0 , \\ S(\psi) & \text{if } \psi_0 \le \psi \le \pi , \end{cases} \qquad (29)$$

and expanding it into a series of Legendre polynomials $P_n(x)$

$$\bar{S}(\psi) = \sum_{n=0}^{\infty}((2N+1)/2)Q_n P_n(\cos\psi) \ , \qquad (30)$$

where Q_n is the Molodenskij coefficients and is defined as

$$Q_n = Q_n(\psi_0) = \int_0^{\pi}\bar{S}(\psi)P_n(\cos\psi)\sin\psi d\psi \ ,$$
$$= \int_{\psi_0}^{\pi} S(\psi)P_n(\cos\psi)\sin\psi d\psi \ , \qquad (31)$$

The rms error caused by neglecting the distant zones is the square root of

$$\overline{\delta N_{TR}{}^2} = (R^2/4\gamma^2)\sum_{n=2}^{\infty} Q_n{}^2 C_n \ , \qquad (32)$$

where the c_n are the anomaly degree variance (Heiskanen and Moritz, 1967).

The error function has two local minima occuring at the zeros of the Stokes kernel. Even if the cap size is extened to sixty degrees, the truncation error still exceeds 10 m.

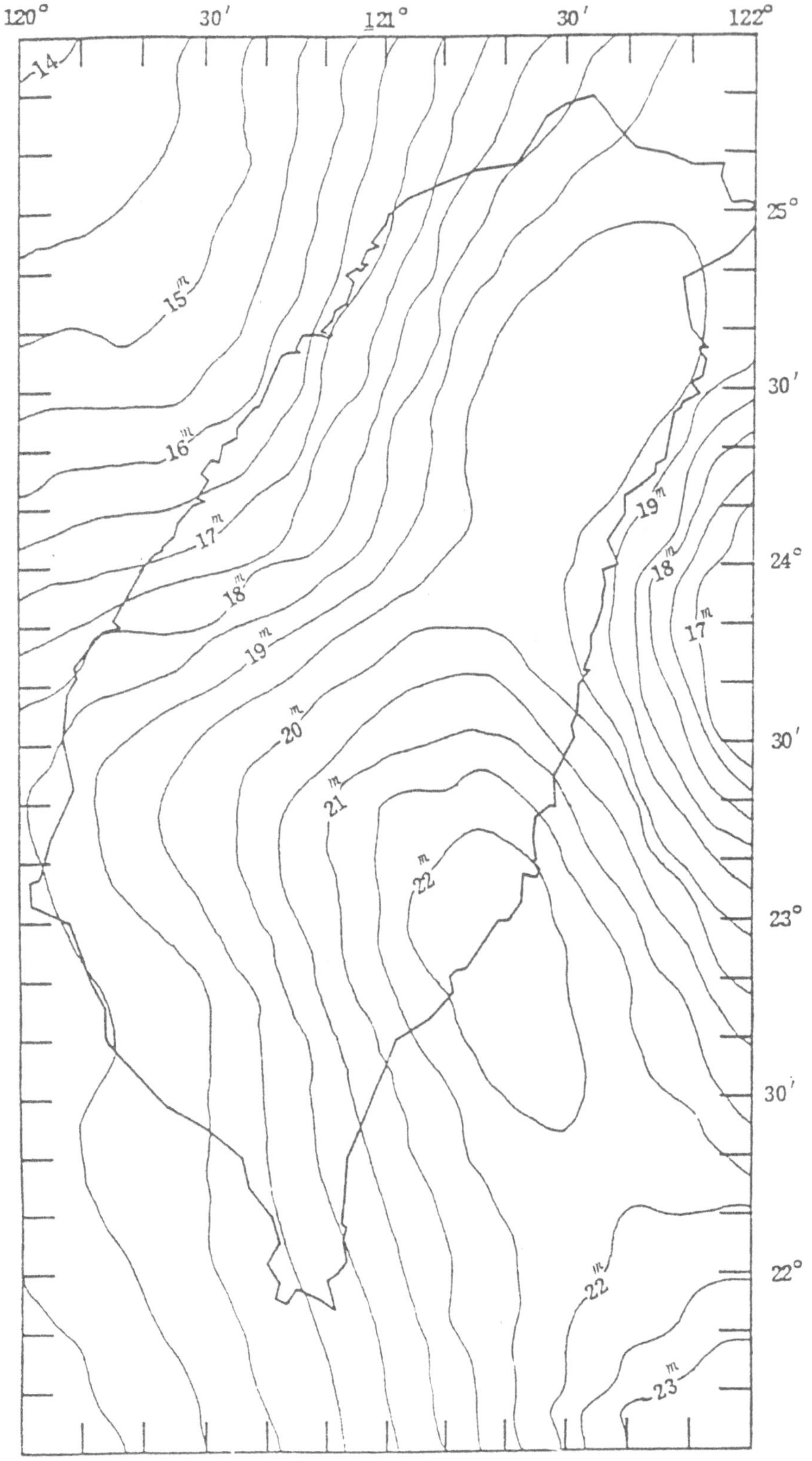

Fig. 1. Medium wavelength contribution of the gravimetric geoidal heights.

Method of Evaluation

Evaluation was in three zones, depanding on distance from the computation point. Appropriate sizes of terrestrial data used for each zone. Table 3.1 contains this information. The computation of this part is done by an integration program CCIT1. This program performs three integration:

1. The innermost zone using the approximation form of the Stokes kernel $S(\psi)$ (Vanicek et al., 1987):

   $$S(\psi) \approx (2/\psi) - 3\ln(\psi/2) - 4 \;, \qquad (33)$$

   The gravity anomaly in this zone is assumed to be representable by a second order algebraic polynomial in x and y according to :

   $$\Delta g^l(x,y) = \alpha_o + \alpha_1 x + \alpha_2 y + \alpha_3 xy + \alpha_4 x^2 + \alpha_5 y^2 \;, \qquad (34)$$

   where $\alpha_i (i=<1,5>)$ are coefficients to be determined as the geometry of the gravitational distribution of point gravity anomalies in the innermost zone allows.

2. The inner zone i ntegration using the modified spheroidal kernel. The final integration procedure is

   $$N_I = (R/4\pi\gamma)\sum_{i=1}^{L} S_M{}^l(\psi_i)\overline{\Delta g_i}{}^l\sigma_i \;, \qquad (35)$$

   $$\text{where } S_M{}^l(\psi) = S(\psi) - \sum_{i=2}^{\ell}((2i+1)/(i+1))P_i(\cos\psi) - \sum_{i=0}^{\ell}((2i+1)/2)t_i(\psi_o)P_i(\cos\psi) \;, \qquad (36)$$

   $\overline{\Delta g_i}{}^l$ assigned to the 10' by 10' blocks.

3. The outer zone integrationusing the modified spheroidal kernel.

   $$N_o = (R/4\pi\gamma)\sum_{i=1}^{M} S_M{}^l(\psi_i)\overline{\Delta g_i}{}^l\sigma_i \;, \qquad (37)$$

   where M is an integer number of 1 by 1 degree cells, $\overline{\Delta g_i}{}^l$ is the 1 by 1 degree mean gravity anomaly.

Using eqns. (34), (36), (37) has produced the short wavelength gravimetric geoid without corrections.

Fig.2 showing with the correction terms for Taiwan area.

CONCLUSIONS

Experiments from the practical computation shows to us the optimal trunction radius for the spheroidal Stokes' integration can save the computation time and maintain the same accuracy.

Combing the short wavelength from the terrestrial data and the medium wavelength from OSU86F geopotential coefficients has some un-expected discoveries. Specially, the result of the gravimetric geoid shows in great detail very localized features. The above mentioned very detailed structure of the gravimetric geoid makes it almost impossible to represent its combination with other data in the form of an algebric surface. A surface representation would either require a tremendous number of base functions or it would smooth the details in the gravimetric geoid. Other kinds of data, e.g., satellite altimetry, and Global Positioning System (GPS), should be included in the solution. It is possible to produce corrections to the gravimetric geoid reflecting the other kinds of data.

We has completed the re-survey of gravity data on the first-order levelling and also in process of GPS observation in whole area. We interested to collect the sea gravity data around Taiwan area (including sea point gravity data and transfered from GeoSat altimeter

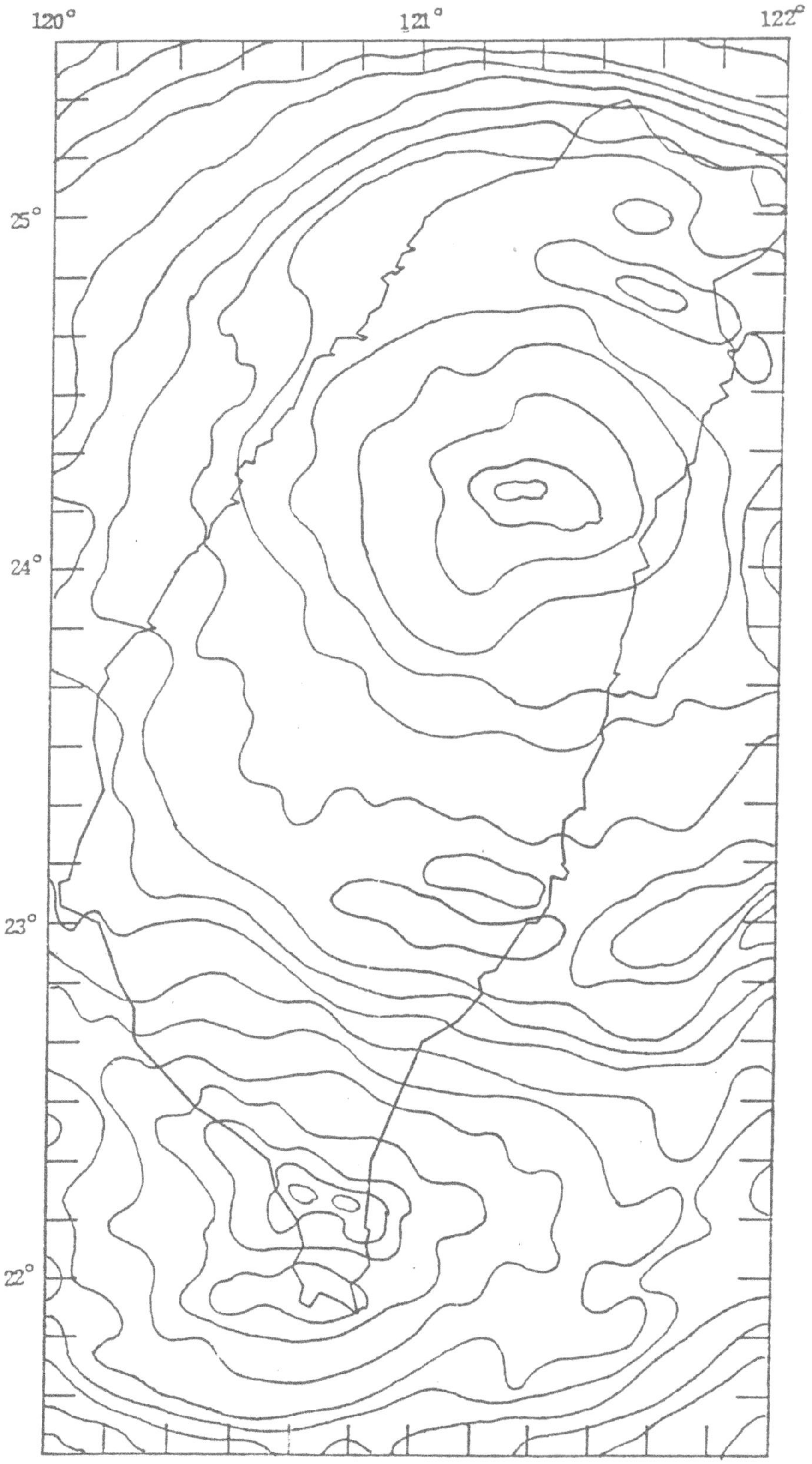

Fig. 2. The gravimetric geoid of Taiwan area.

data) with respect to the other new data sets of the whole area enhanceing the accuracy of gravimetric geoid in Taiwan area in the future.

Acknowledgement. This research was supported by the National Science Council, Republic of China under Contract No. NSC-78-0410-E-014-02. We would like to thank Prof. R.H. Rapp (OSU) for contributing to us the valuable data.

REFERENCES

Ecker, E., and Mittermayer, E. (1969). Gravity corrections for the influence of the atmosphere, *Boll. Geofisica Teor. Appl.*, **11**, pp.70-79.

Engelis, T., Rapp, R.H., and Tscherning, C.C. (1984). The precise computation of geoid undulation differences with comparison the results obtained from the GPS, *Geophys.Res. Lett.* 1, No.9, pp.821-824.

Heiskanen, W. and Moritz, H. (1967). *Physical geodesy*, Freeman.

IAG (1971). Geodetic reference system, Special Publication No.3, Central Bureau of the IAG, Paris, France.

Kassim, F.A. (1980). An evaluation of three techniques for the prediction of gravity in Canada, *Dept. of Sur. Eng. Tech. Rep.* **73**, Univ. of New Brunswick, Fredericton.

Lambert, W., and Darling, F.W. (1936). Tables for determining the form of the geoid and its indirect effect on the gravity, *Special Publication* No.199, p.114, U.S. Coast and Geodetic Survey, U.S. Supt. of Doc. Washington, D.C.

Lelgemann, D. (1970). Untersuchungen Zue Einir Genaueren Losung des Problem von Stokes, Sec.C 155, Dtsch.Geod.Komm., Munich, FRG.

Marsh, J.G., and Chang, E.S. (1976). Detailed gravimetric geoid confirmation of sea surface topography detected by the Skylab S-193 altimeter in the Atlantic ocean, *Bulletin Geodesique*, **50**, No.3, pp.291-299.

Mather, R.S. (1973). A Solution of the Geodetic Boundary Value Problem to Order e^3, Goddard Space Flight Center Preprint X-592-73-11, Greenbelt, U.S.A.

Molodenskij, M.S., Eremeev, V.F., and Yurkina, M.I. (1962). Methods for study of the external gravitational field and figure of the earth, Translated from Russian by the Israel Program for Scientific Translations for the Office of Technocal Service, U.S. Department of Commerce, Washington, D.C.

Moritz, H. (1974). Presise gravimetric geodesy, *Dept. of Geod. Sci. Rep.* **213**, Ohio State Univ., Columbus.

Moritz, H. (1980). *Advanced physical geodesy*, Abacus, Kent, England.

Rapp, R.H.(1980). A comparison as altimeter and gravimetric geoids in the Tonga Trench and Indian ocean areas, *Bulletin Geodesique*, **54**, pp.1149-163.

Rapp, R.H. (1981). The earth's gravity to degree and order 180 using altimeter data, terrestrial gravity data, and other data, *Dept. of Geod.Sci. Rep.* **322**, Ohio State Univ., Columbus.

Rapp,R.H. (1983). The determination of geoid undulations and gravity anomalies from Seasat altimeter data, *J. Geophys. Res.*, **88**, No.C3, pp.1552-1562.

Rapp, R.H. and J.Y. Cruz (1986). Spherical harmonic expansions of the earth's Gravitational potential to degree 360 using30' mean anomalies. *Dept. of Geod. Sci. Rep.* **376**, Ohio State Univ., Columbus.

Rapp, R.H., and Rummel, R. (1975). Methods for the computation of detailed geoid and their Accuracy, *Dept. of Geod. Sci. Rep.* **233**, Ohio State Univ., Columbus.

Vanicek, P. and John, S.(1983). Evaluation of geoid solutions for Canada using difference kinds od data, *Presented at t he XVIII General Assembly, I nternational Union of Geodesy and Geophysics*, Hamburg, FRG.

Vanicek, P., Kleusberg, A., Chang, R.G., Fashir, H., Christou, N., Hofman, M., Kling, T., and Arsenault, T. (1987). The Canadian geoid, *Dept. of Sur. Engi. Rep.* **127**, Univ. of New Brunswick, Fredericton.

Vanicek, P. and Krakiwsky, E. (1982). *Geodesy: The Concepts*, North-Holland Publishing Company, Holland.

GEOIDAL MAP IMPROVEMENT IN BRAZIL

Denizar Blitzkow
IAG - USP
Caixa postal 30627 - 01051 São Paulo - Brazil.

Luiz Paulo Souto Fortes
René Armando Zepeda Godoy
IBGE
Av. Brasil, 15671 - 21241 Rio de Janeiro - Brazil.

ABSTRACT

A very great effort has been done recently by many different organizations in Brazil in order to improve the gravity network. The consequence has been an increasing number of blocks of $1^0 x 1^0$ where mean gravity anomaly has been possible to estimate. In the other hand, IBGE has increased the number of Doppler observations in the levelling network and improved the computations of the observations with new programs. This paper presents the situation of mean gravity anomalies now available and the results of the comparison between the Doppler undulations and the recent GEM-T1 model. Finally, a geoidal map using 301 doppler undulations and the spheroidal heights derived from GEM-T1 is presented.

INTRODUCTION

Most of the gravity data now available in Brazil belong to the Brazilian Company of Petroleum (PETROBRAS). The gravity network stablished by PETROBRAS has very often a density of a few hundred meters between points, but restricted to areas more reliable for petroleum exploration, in general sedimentary basins. This characteristic left many blank areas to be filled by other organizations. The Observatório Nacional (ON) had been, for many years, the only official organization worried about the gaps. Recently, different universities joined themselves to ON in order to improve as much as possible the gravity network in Brazil. Some names have to be mentioned: Federal University of Paraná, Federal University of Brasília, Federal University of Pará, Federal

University of Rio Grande do Norte, Companhia de Pesquisas de Recursos Minerais (CPRM) and the University of São Paulo. Particularly the last one, through the Institute of Astronomy and Geophysics (IAG - USP), has been involved with many gravity activities, in particular, measurements on the first order levelling network, as a consequence of an existing agreement between IAG and Fundação Instituto Braileiro de Geografia e Estatística (IBGE). These efforts became possible the computation of mean gravity anomalies in blocks of $1^0 \times 1^0$ by [SÁ, 1989] whose distribution is presented in the Map 1. These gravity data will certainly improve the computation of the geoid in the near future.

In the other hand, IBGE, responsible for the geodetic networks, has devoted the last few years in the stablishement of Doppler points in the first order levelling network. The precise ephemeris available through facilities of IAGS-DMA, allowed the computation of the Doppler coordinates in the NSWC-9Z2 reference system until December 31, 1986 and since then in the WGS-84. Most of the observations are being reduced now with the GEODOP-V program, kindly made available by Dr. Kouba. In Brazil there are some Doppler coordinates related to the broadcast ephemeris. Nevertheless, for this paper only coordinates related to the precise ephemeris have been used.

RESULTS

Many different comparisons have been done between Doppler undulations and the former gravitational models in Brazil [BLITZKOW & SÁ,1983], [FORTES & BLITZKOW,1988], [BLITZKOW et al.,1987].

Now a total of 335 Doppler undulations have been compared with the GEM-T1 model. After a preliminary analysis of the differences, a total of 34 points reached values between 5m to 12m and have been removed. The problem of those points seems to come from the old program used by IBGE to reduce the Doppler observations. They will be now reduced with the GEODOP-V. In this way, 301 values have been compared with the model. An RMS difference of 2.3 m has been found. This value shows an agreement between the model and the Doppler better than any model already available. Fig. 1 shows the histogram of the differences. Using the mentioned Doppler undulations and the spheroidal heights derived from GEM-T1 in the blocks of $0.5^0 \times 0.5^0$ without Doppler, a geoidal chart referred to SAD-69 (South American Datum 1969) is presented (Map 2). The translation parameters used to refer geocentric values to local are: [FORTES et al., 1989]

$$TX = 66.87 \text{ m} + 0.43 \text{ m}$$

$$TY = -04.37 \text{ m} + 0.44 \text{ m}$$

TZ= 38.52 m + 0.40 m

The following corrections have been added previously to the NSWC-9Z2 coordinates according to DMA (1987):

TZ= 4.5 m

w3= -0.814"

K= -0.6 PPM

to turn them to geocentric.

The original scale of the map is 1:10 000 000 in a policonic projection. The interpolation was done with Krigging algorithm in order to have a regular grid of 0.025 m in the mentioned scale. It has been fixed a maximum of 4 points in the neighbourhood of the grid point for the interpolation.

FINAL CONSIDERATIONS

Based on the RMS difference and on the maximum difference between the Doppler and the model it is expected an absolute error of 3 meters and a relative error of 0.2 m/20km for the geoidal heights derived from the map where doppler points exist.

After new Doppler points have been added to the file, the GEODOP-V program has been implemented and the WGS-84 reference system has been introduced, this is a first analysis of the Doppler undulations. In the next few months, after the process of the observations with GEODOP-V will be completed, new results may lead to other conclusions and the present geoidal map may change a little.

In the other hand, the mean gravity anomalies now available, as mentioned before, will certainly improve the geoidal heights at least in some areas of the country. For this purpose, computations are going on in order to use the modified Stokes integral and the least squares collocation. A comparison with the two methods will certainly be useful.

ACKNOWLEDGMENT

The opportunity to participate of the General Meeting of AIG is possible through funds kindly provided by Fundaçao de Amparo à Pesquisa do Estado de São Paulo (FAPESP) and by the International Association of Geodesy. Many profitable discussions with Nelsi Côgo de Sá are always appreciated.

REFERENCES

BLITZKOW D. and SÁ N. C. DE (1983). As alturas geoidais Doppler e a separação elipsóide-esferóide de referência. Revista Brasileira de Geofísica, 2(1), pp. 19-24.

BLITZKOW D.; SÁ N. C. DE AND FORTES L. P. S. (1987). Efforts toward a geoidal chart of Brazil referred to SAD-69. XIXth. General Assembly of IUGG, Vancouver, August 9-22.

DEFFENSE MAPPING AGENCY (DMA) (1987). World Geodetic System 1984. Its definition and relationships with local systems. Report 8350.2.

FORTES L. P. S. AND BLITZKOW D. (1988). As alturas geoidais Doppler e os modelos de grau elevado do potencial gravitacional. Cadernos de Geociências, Fundação IBGE, no. 1, pp:23-27.

FORTES L. P. S.; CAGNIN I. F.; GODOY R. A. Z. e BLITZKOW D. (1989). Determinação dos parâmetros de transformação entre os sistemas NWL-10D, NSWC-9Z2 e o SAD-69. Anais do XIV Congresso Brasileiro de Cartografia, vol.I, pp: 157-165.

SÁ N. C. de (1989). Modelos estatísticos e representação integrada do campo de gravidade no Brasil. Doctor Thesis presented to the Department of Geophysics, University of São Paulo.

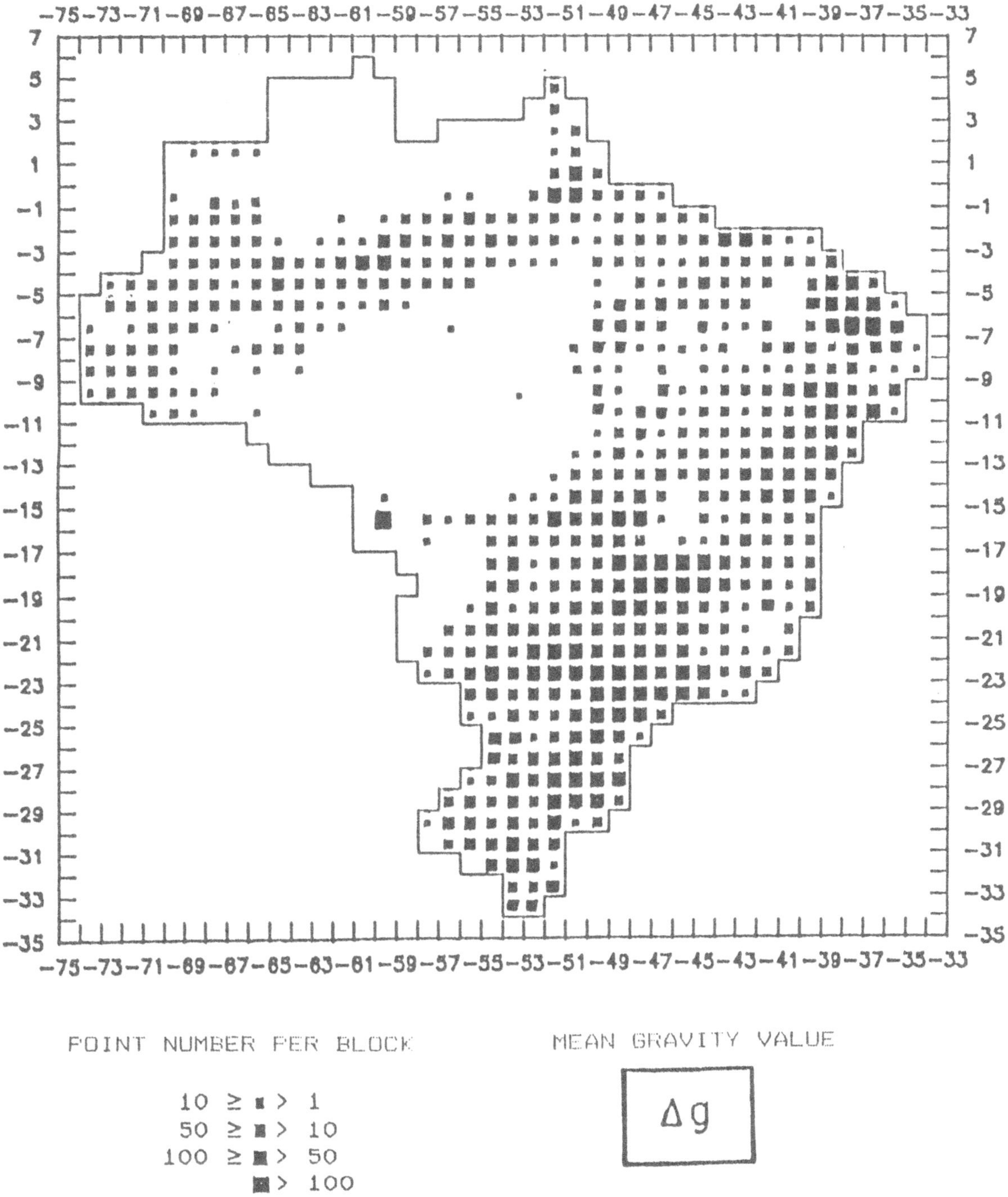

MAP 1 - MEAN GRAVITY DISTRIBUTION.

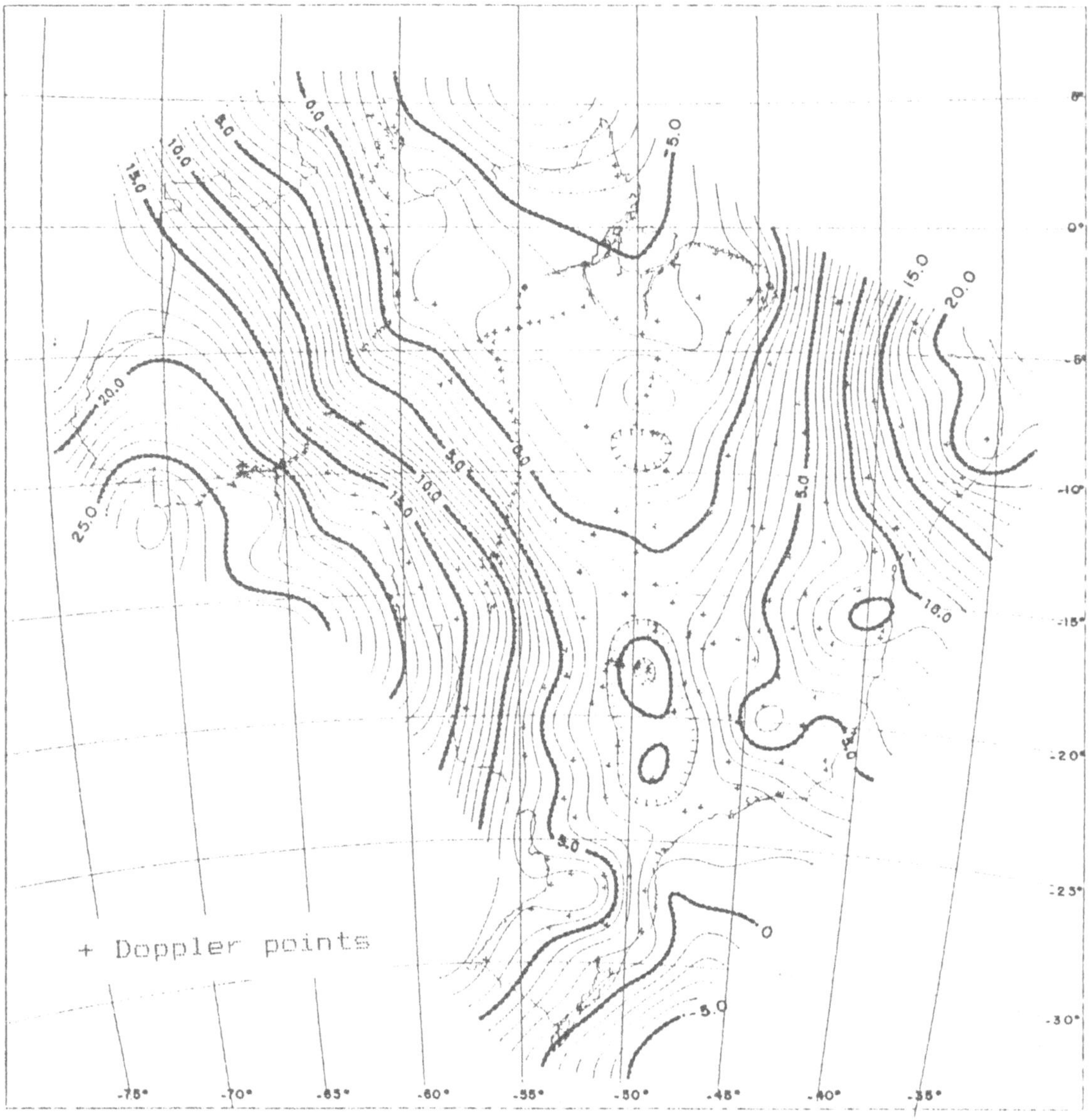

MAP 2 - GEOIDAL MAP OF BRAZIL (SAD-69)-PRELIMINARY.

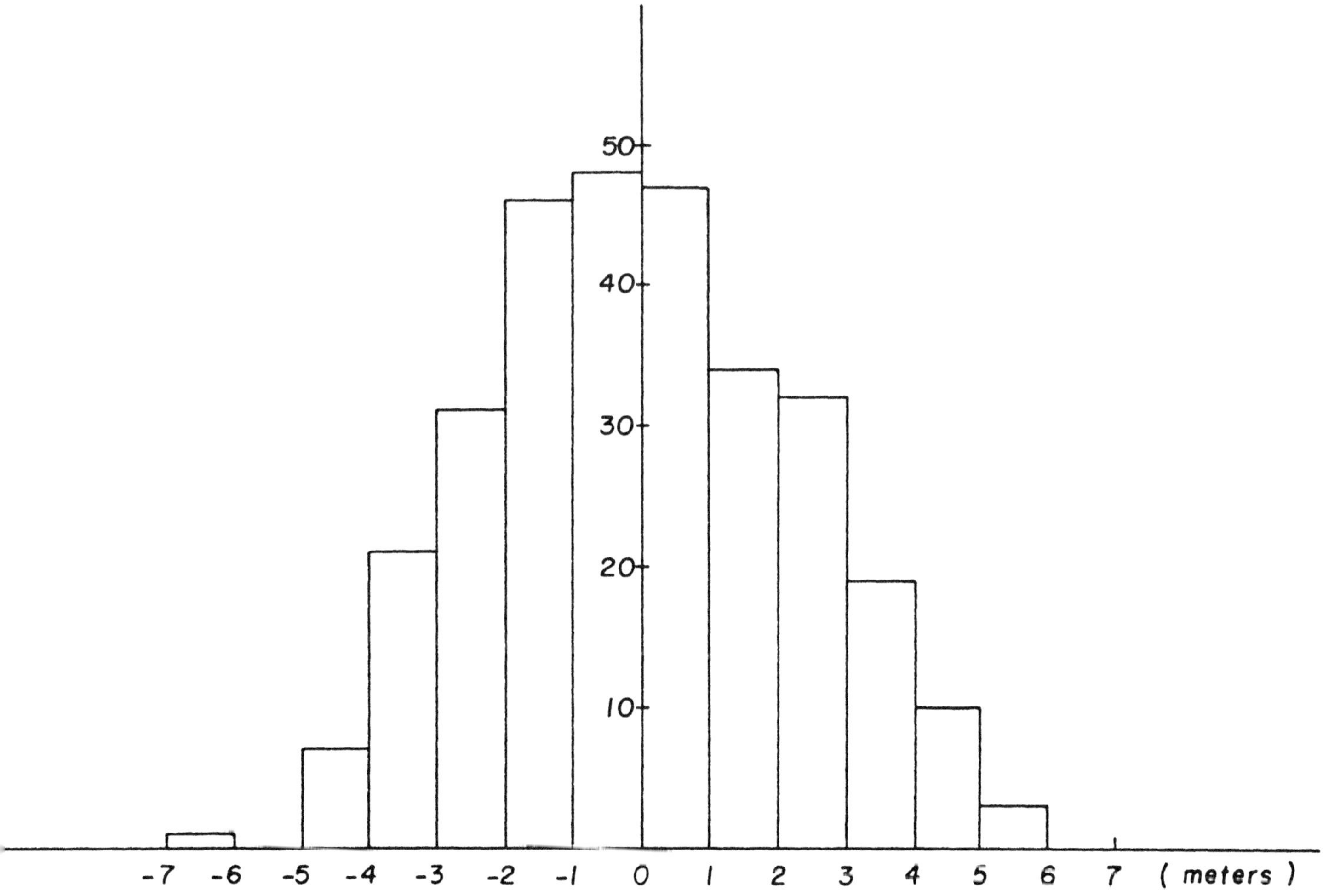

Fig. 1 Histogram of the differences

CONTROL AND REFINEMENT OF THE AUSTRIAN GEOID USING GPS

E. Erker
Federal Bureau of Surveying and Metrology
Schiffamtsgasse 1 - 3
A - 1025 Vienna, Austria

H. Sünkel
Institute of Mathematical Geodesy
Graz University of Technology
Technikerstraße 4
A - 8010 Graz, Austria

THE GEOID IN AUSTRIA

Historical Background

The local geoid determination for Austria has a history of more than three decades. In 1953 Litschauer made a first and quite successful attempt of an astrogeodetic geoid determination for Austria, based on about a hundred vertical deflections which corresponds to about 1 point per 1000 km^2. Without the consideration of the topographic effect (which could not be computed at that time) he achieved a geoid solution with $\pm$ 9 cm rms error over a 10 km sidelength.

After a large number of vertical deflections had been observed mainly in the late seventies, a detailed geoid determination seemed feasible, at least for 2/3 of the Austrian territory. Based on a pilot study (Sünkel, 1982) a first detailed astrogeodetic geoid determination was pursued independently by the Federal Bureau of Surveying and Metrology in Vienna (Erker, 1983) and by the Institute of Theoretical Geodesy of the Graz University of Technology (Sünkel, 1983). In both solutions over 500 vertical deflections had been used. The Vienna solution relied on Helmert's classical method of astrogeodetic geoid determination, supplemented by a topographic remove-restore procedure. In the Graz solution least squares collocation was used, supplemented by a topographic-isostatic remove-restore technique and using the OSU 81 Earth model as a global reference. These two solutions turned out to be about 10 times as accurate as the early solution of Litschauer.

After vertical deflection measurements had been completed for the western part of Austria, a new complete geoid solution suggested itself. Using basically the same concepts as in 1983, two independent geoid solutions were performed in Vienna and in Graz in 1987 (Erker, 1987; Sünkel, 1987). The difference of the two solutions was found to be of the order of 20 cm rms (Erker, 1987).

Source and Effect Data

In applied (geo)sciences it has become customary to discriminate between source data and effect data. In the context of the (Earth's) gravity field the source is the mass density distribution inside the Earth's surface, the effect is the resulting gravity field. By source data we understand mass density data which are available in terms of digital terrain models (DTM) and digital surface density models (DDM). For the above referenced solution the following source data were at our disposal:

- DTM for Austria, $11.25'' \times 18.75''$
- DTM global, TUG87, $5' \times 5', 1^o \times 1^o$
- DDM for Austria, $1.5' \times 2.5'$
- DDM Eastern Alps, $3' \times 5'$

By effect data we understand gravity field data in the proper sense such as gravity anomalies, vertical deflections, etc. For the above referenced solution the following effect data were available:

- 26000 gravity data
- 683 vertical deflections
- 41 Doppler points + preliminary orthometric heights
- OSU 81 (86) Earth model

It should be mentioned that the gravity data have not been used explicitly in the geoid solutions because of inhomogeneous data coverage and because of missing access to data in the countries surrounding Austria clockwise from north to south. Gravity data have rather been used indirectly for the estimation of the covariance funtion which underlies the collocation solution (Sünkel, 1987). Erker (1987a) has used gravity anomalies for gravimetric levelling tests in the central part of Austria.

The Solution

Numerical experiments in several test areas on various parts of the Earth during the last decade have clearly demonstrated that a topographic-isostatic remove-restore process is superior to any other gravity field estimation technique. Therefore, such a procedure suggested itself. Meanwhile this procedure has become so widely known and generally accepted that we are only presenting a kind of flowchart diagram.

Denoting the Earth model (like OSU 81) by EM, topography/isostasy by TI, the Helmert astrogeodetic solution by HA and a least squares collocation solution by LSC , the entire solution process is as follows:

$\delta l = l - l_o$ ... remove process

$\delta\hat{T} = L^{-}\,\delta l$ ... HA or LSC solution

$\hat{T} = T_o + \delta\hat{T}$ ... restore process

with

T ... Gravity potential

T_o ... model gravity potential (EM, TI)

$\delta T = T - T_o$ ... residual potential

$l = LT$ ... gravity field data

$l_o = LT_o$ ... model effect

Having obtained an (optimal) estimate for the residual potential $\hat{T}$, we may predict any other gravity field quantity by LSC,

$L_s\hat{T}$ ... predicted quantity

Results

The gravity field solution 1987 yielded the following prediction accuracies:

Vertical deflection:	0.8 " (verified)
Absolute gravity:	1.5 mGal (verified)
Geoidal height:	5 cm / 100 km (locally verified)

THE GPS CAMPAIGN AGEDEN

In 1987 it was decided to run an Austrian-wide GPS campaign with the main purpose to establish a regional GPS network in several geodynamically active areas in Austria. The campaign was part of the WEGENER/MEDLAS-project. It was supposed to be a basis for re-measurements in regular time intervals to detect recent crustal movements.

Three active areas had been identified:

- Periadriatic boundary (southern part)
- Northern Tauern boundary (central part)
- Diendorf tectonic disturbance (northern part)

During 8 days (day 255 - 262, 1987) 47 GPS points have been observed using 8 TI 4100 receivers. Keeping the orbit and one reference point fixed, using the Saastamoinen tropospheric model, assuming an a priori rms error of the single differences of $\pm$ 1 cm, and performing a least squares adjustment based on L3, the following results have been achieved (Hofmann et al., 1989):

- daily solution (DS): $\pm 11 - 23mm$
- complete solution (CS): $\pm 4 - 45mm(X, Y, Z)$
- transformation of DS on CS: $\pm 1ppm$
- transformation of CS on "Gebrauchsnetz": $\pm 50cm$ horizontal, $\pm 20cm$ vertical

From the results of this AGEDEN GPS - campaign and its comparison with the "Gebrauchsnetz" we may draw the following conclusions:

- GPS results of a single campaign yield $\pm 1ppm$
- A comparison of GPS coordinates with coordinates of the "Gebrauchsnetz" makes regionally little sense because of inhomogeneities of the latter

COMPARISON OF HEIGHTS

The new system of orthometric heights for Austria is being established now, but is not yet completed. However, for 27 of the 47 AGEDEN points preliminary geopotential numbers are already available. Orthometric heights have been determined at these points using the simple empirical, but sufficiently accurate approximate relationship due to Bretterbauer (1987), which is accurate to about $\pm$ 2 mm rms,

$$H = \frac{C}{\bar{g}}$$

$$H \doteq \frac{C}{g} - 34 \cdot 10^{-9} H^2 \, [m]$$

(H denotes orthometric height, C geopotential number, g observed absolute surface gravity, $\bar{g}$ mean gravity along the plumb line between the surface point and the corresponding point on the geoid.) The following results have been obtained for the comparison of GPS-derived ellipsoidal heights versus ellipsoidal heights derived from geoid heights and (these preliminary) orthometric heights with only small differences in minimum, maximum and rms error between the Vienna and the Graz solution:

GPS versus terrestrially determined heights Unit: 1 cm			
Solution	*min.*	*max.*	*rms*
Geoid + spheroidal height	-30	+22	13
Geoid + orthometric height	-24	+19	9

Fig. 1 shows the error bars of the 27 height differences both for the case of spheroidal (normal-orthometric) and for orthometric heights.

The discrepancies between heights are due to errors in the geoid heights, errors in the orthometric heights, and errors in the GPS-derived heights; (the latter are particularly visible in well-known insufficiencies of the GPS-solution in the northern part of Austria.) When obvious gross errors are excluded in our solution, the rms error of height differences drops to an astonishing value of only ± 4 cm which can be termed excellent considering the mountainous character of most of the Austrian terrain.

In the near future an improved geoid and adjusted orthometric heights are expected to bring the differences further down by a factor of probably 2; a recently planned large-scale GPS-campaign for Austria will significantly contribute to an improvement of the geoid, in particular in those areas which are not yet sufficiently covered by gravity data; and in the far future we may even envision to replace the cumbersome levelling by a combination of GPS-provided geometric heights and gravity field provided geoidal heights.

REFERENCES

[1] Bretterbauer, K. (1987). Approximative Meereshöhen, *ÖZfV* 1, 10 - 13.

[2] Erker, E. (1983). Das Geoid in Österreich, Berechnungen des Bundesamtes für Eich- und Vermessungswesen, In: Das Geoid in Österreich, *Geodätische Arbeiten für die Internationale Erdmessung*, Neue Folge, III, 89 – 115, Graz.

[3] Erker, E. (1987). The Austrian geoid - local geoid determination using modified conservative algorithms, In: The Gravity Field in Austria, *Geodätische Arbeiten für die Internationale Erdmessung*, Neue Folge, IV,19 – 46, Graz.

[4] Erker, E. (1987a). Das österreichische Geoid - Die vollständige Lösung 1987. *ÖZfV* 4, 202 – 206.

[5] Hofmann-Wellenhof, B., G. Kirchner, H. Lichtenegger, H. Moritz, P. Pesec, K. Rinner, G. Stangl, and H. Sünkel (1989). Österreichische Beiträge zum WEGENER/MEDLAS-Projekt. *Mitteilungen der geodätischen Institute der Technischen Universität Graz*, 65, Graz.

[6] Litschauer, J. (1953). Zur Frage der Geoidgestalt in Österreich, *ÖZfV* 6, 161 – 172.

[7] Sünkel, H. (1982). Das Geoid im Testnetz Steiermark. In: Geodaesia Universalis, Festschrift Karl Rinner zum 70. Geburtstag, *Mitteilungen der geodätischen Institute der Technischen Universität Graz*, 40, 331 – 345, Graz.

[8] Sünkel, H. (1983). Geoidbestimmung, Berechnungen an der TU Graz, 2. Teil, In: Das Geoid in Österreich, *Geodätische Arbeiten für die Internationale Erdmessung*, Neue Folge, III, 89 – 115, Graz.

[9] Sünkel, H., N. Bartelme, H. Fuchs, M. Hanafy, W.-D. Schuh, and M. Wieser (1987). The Gravity Field in Austria. *Geodätische Arbeiten für die Internationale Erdmessung*, Neue Folge, IV, 47 – 75, Graz.

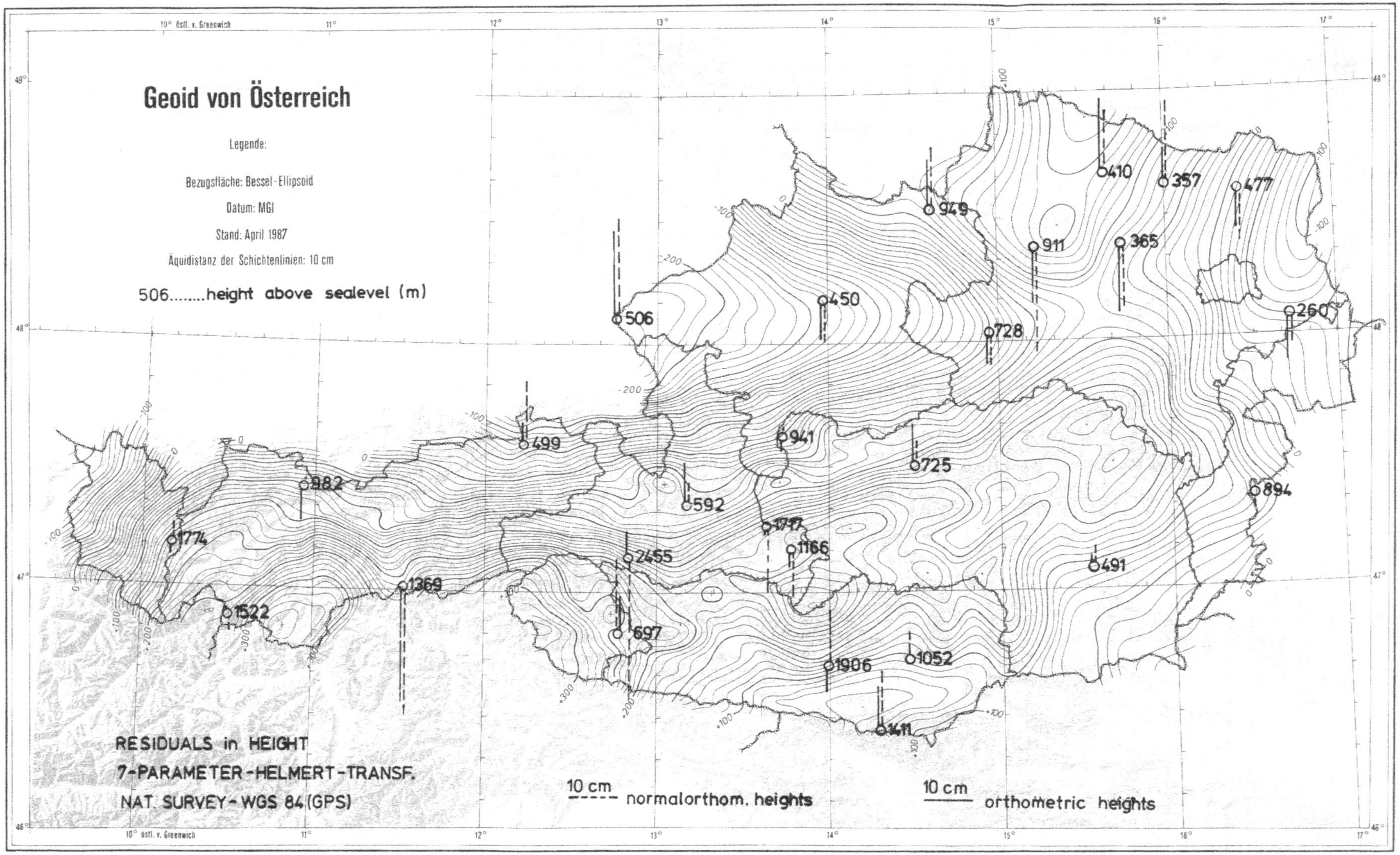

Fig. 1. Discrepancies between GPS-derived and terrestrially determined heights

ALTIMETRY-GRAVIMETRY PROBLEM: AN EXAMPLE

R. Barzaghi, M. Brovelli
Ist. di Topografia, Fotogrammetria e Geofisica
Politecnico di Milano - Milano (Italy)

F. Sacerdote
Dip. di Matematica
Università di Pisa - Pisa (Italy)

ABSTRACT

The topographic surface of the sea can be determined by subtracting the geoid from the geometric height of the stationary sea surface. This latter can be obtained from altimetry measurements; the former can be derived from marine gravity measurements, treated in the form of gravity disturbances.
When a closed sea is treated, the gravity anomalies on land must be combined to the gravity disturbances to give the so called altimetry gravimetry problem, which in spherical approximation is translated into a suitable integral equation.
A relatively small set of data of this type in the Mediterranean Sea is used as an example.
The aim is to compare this procedure with a classical collocation solution: this has been done specifically on the land area. Since the same corrections (e.g. topographic corrections) apply to both methods, the comparison has been performed only for the part of the signal which is usually estimated by collocation and in particular Bouguer anomalies on land have been utilized as input data .

1. FORMULATION OF THE PROBLEM

The altimetry-gravimetry boundary value problems were first formulated in the early eighties (Sansò 1981, Holota 1982) in order to take into account satellite altimetry data on the oceans, as well as gravity measurements for the determination of the anomalous gravity potential of the earth. Information on sea surface topography can be used to obtain directly the geoid height on the surface of the oceans, or, in alternative, marine gravimetry leads to the determination of gravity disturbances on sea. Two different types of altimetry-gravimetry problems are consequently formulated (for details, see for example (Sacerdote & Sansò 1987)).
The second problem, which in its linearized form and in spherical approximation can be written as

$$\begin{cases} \nabla^2 T = 0 & \text{outside } \Sigma \\ -\dfrac{\partial T}{\partial r} - \dfrac{2}{r} T = \Delta g & \text{on } \Sigma_L \text{ ("land")} \\ -\dfrac{\partial T}{\partial r} = \delta g & \text{on } \Sigma_S \text{ ("sea")} \\ T = O(1/r) & (\Sigma = \Sigma_L \cup \Sigma_S) \end{cases} ,$$

can be easily transformed, if gravity anomaly data are reduced to sea level, by using Hotine's formula

$$T(P) = \frac{R}{4\pi} \int_\sigma H(\psi_{PQ}) \left(-\frac{\partial T}{\partial r}\right)(Q)\, d\sigma_Q \tag{1.2}$$

where σ is the unit sphere, $H(\psi_{PQ})$ is the well-known Hotine kernel (Sacerdote & Sansò 1983). The result is

$$T(P) - \frac{1}{2\pi} \int_{\sigma_L} H(\psi_{PQ}) T(Q)\, d\sigma_Q = \frac{R}{4\pi} \int_\sigma H(\psi_{PQ}) f(Q)\, d\sigma_Q \tag{1.3}$$

where $\sigma_L(\sigma_S)$ is the projection of $\Sigma_L(\Sigma_S)$ onto σ ;

$$f(Q) = \begin{cases} \Delta g & \text{if } Q \in \sigma_L \\ \delta g & \text{if } Q \in \sigma_S \end{cases}$$

(1.3) is first used as an integral equation to obtain T(P) for $P \in \Sigma_L$; then for $P \in \Sigma_S$, T(P) is directly calculed using (1.3) as a simple integral formula.

The numerical treatment of (1.3) over the whole sphere cannot be carried out directly owing to the enormous amount of data required; on the other hand, by virtue of the singularity of $H(\psi)$ for small ψ and of its fast decreasing with growing ψ, the essential contribution to the integrals in (1.3) comes from a small area around the computation point P, so that (1.3) can be approximated by replacing the data in regions far from P with values computed from a geopotential model, according to a procedure described in (Sansò & Stock 1985). Such computed values can be any geodetic quantities, for example gravity disturbances, as if the whole earth's surface, but for a small area around the computation point, were covered by oceans. Consenquently the integration domain σ in (1.3) is split into a cap σ_C around P and its complement σ_R; one obtains

$$T(P) - \frac{1}{2\pi}\int_{\sigma_{LC}} H(\psi_{PQ})T(Q)\, d\sigma_Q = \frac{R}{4\pi}\int_{\sigma_C} H(\psi_{PQ})f(Q)\, d\sigma_Q + \tag{1.5}$$

$$+ \frac{R}{4\pi}\int_{\sigma_R} H(\psi_{PQ})\left(-\frac{\partial\tilde{T}}{\partial r}\right)(Q)\, d\sigma_Q$$

where:

$\sigma_{LC} = \sigma_L \cap \sigma_C$; $\tilde{T}$ is the approximate model adopted for T.

This "localization" procedure enables to compute the anomalous potential and the geoid in a restricted region from data distributed on a slightly larger region surronding it; its application is particularly fit for islands of medium size, as can be found for example in the Mediterranean (Sicily, Sardinia, Cyprus, etc.).
Finally, if one subtracts from (1.5) the identity

$$\tilde{T}(P) - \frac{1}{2\pi}\int_{\sigma_{LC}} H(\psi_{PQ})\tilde{T}(Q)\, d\sigma_Q = \frac{R}{4\pi}\int_{\sigma_C} H(\psi_{PQ})\tilde{f}(Q)\, d\sigma_Q + \tag{1.6}$$

$$+ \frac{R}{4\pi}\int_{\sigma_R} H(\psi_{PQ})\left(-\frac{\partial\tilde{T}}{\partial r}\right)(Q)\, d\sigma_Q \quad ,$$

that must be fulfilled if $\tilde{f}(Q)$, in analogy with (1.4), represents gravity anomalies or disturbances computed from $\tilde{T}$, one obtains

$$\delta T(P) - \frac{1}{2\pi}\int_{\sigma_{LC}} H(\psi_{PQ})\delta T(Q)\, d\sigma_Q = \frac{R}{4\pi}\int_{\sigma_C} H(\psi_{PQ})\delta f(Q)\, d\sigma_Q \tag{1.7}$$

where:

$\delta T = T - \tilde{T}$; $\delta f = f - \tilde{f}$.

The amplitude of the cap σ_C must be chosen as small as possible, provided that the error due to the introduction of $\tilde{T}$ outside σ_C be comparable with the numerical discretization error.
Numerical simulations are described in (Sansò & Stock 1985) for simplified geometry of land and sea distribution. The present paper represents the first attempt to approach real data following this procedure, with all the practical problems involved in this area. In particular as a result no specific computation is optimized and the results should be considered as absolutely preliminary.

2. NUMERICAL PROCEDURE AND RESULTS

The region chosen for the numerical experiment was Sicily, for which a reasonable amount of data was available:
1111 altimetry measurements from SEASAT (on 39 tracks) and 2304 from GEOSAT (on 78 tracks);
1762 5'x5' gravity anomalies on sea from DGGG80 (University of Tessaloniki);
1075 gravity anomalies on land of the Italian gravity file.
A complete Bouguer reduction was applied to land data.
As for the treatment of marine gravity, our purpose being the computation of $\delta g = g(P) - \gamma(P)$, we must be able to find $\gamma(P)$, i.e. essentially the height of P on the geoid. This is what can be derived from altimetry, however after long wavelength radial orbit error has been eliminated (Schrama 1989). This is usually done by cross-over adjustment, assuming some polynomial form of the error: often a simple linear expression, which seems suitable in our case, with very short arcs.
What we have really done is an adjustment of each altimetric arc modified with a function linear in time to the corresponding geoid profile, computed by collocation from the marine gravimetry, used in the form of gravity anomalies $\Delta g = g(P) - \gamma(Q)$, with Q the foot of the vertical from P on the ellipsoid. In this way still a certain mixture of the SST and geoidal heights is possible, that in any case is intrinsic to the problem. In any way this procedure was successfull at least in reducing the cross-over residuals (Figg. 1, 2).
The sea surface so computed, has then been interpolated with the spline method to supply a set of heights on the geoid on which the gravity was known: in this way the gravity disturbance set has been constructed.
The global potential model GPM2 (Wentzel 1985) was used to compute $\tilde{T}$ and the related geodetic quantities.
The chosen amplitude of the cap σ_C was 1°. This choice proved to be non optimal, but it was the best possible to exploit the region in which data were available.
Equation (1.7) was discretized on a 3'x3' grid containing 6000 points.
The geodetic quantities in grid points were predicted from data by spline interpolation; the corresponding model quantities were computed by GEOCOL86; 1105 grid points on land were chosen as computation points (Fig. 3).
The discretized equation has the form:

$$\delta T_i - \frac{1}{2\pi} \sum_j \bar{H}_{ij} \, \delta T_j = \frac{R}{4\pi} \sum_k \bar{H}_{ik} \, \delta f_k \qquad (2.1)$$

where:

$\delta T_i = \delta T(P_i)$

$\delta f_k = \delta f(P_k)$

$$\bar{H}_{ik} = \begin{cases} H(\psi_{P_iP_k})\ \mu(\sigma_k) & \text{if } \psi_{P_iP_k} \le \bar{\psi} \\ \displaystyle\int_{\sigma_i} H(\psi_{P_iQ})\ d\sigma_Q & \text{if } k = i \\ 0 & \text{if } \psi_{P_iP_k} > \bar{\psi} \end{cases}$$

with:

P_i computation point

P_k any grid point

$\mu(\sigma_k)$ area sorrounding a grid element

$\bar{\psi}$ amplitude of σ_C

σ_i grid surface elements

The sums with index j are over computation points; the sums with index k are over grid points. Hence the dimension of the linear system matrix, which is symmetric, is 1105x1105, while a large amount of computations must be carried out in order to determine the known terms.

The reason for separately defining the diagonal elements lies in the singularity of the Hotine kernel, which is a weak singularity, with finite surface integral. The computation of these elements can be simplified by approximating the integrals over the grid elements with integrals over equivalent caps centred in computation points, that can be determined analitically and are expressed by a simple formula.

The system was solved by Cholesky decomposition. The geoid heights obtained from the solution via Bruns formula (Fig. 4) were compared with the collocation solution from gravity data (Fig. 5).

The agreement in the general behaviour of the two solutions, as can be seen (Fig. 6) from a first rough look at the figures, appears quite good, but for the northern strip, in which, however, data are less reliable.

Yet, from a tighter numerical analysis it comes out that the mean of the differences is 88 cm and the r.m.s. 55 cm contrary to the visual intuition. This gap does not decrease significantly if the northern strip is skipped, and that's because it is not due to local irregularities, but to an overall bias which can be rather interpreted as a long wavelength error which is not sufficiently controlled by our data set with its limited window.

In spite of this fact, certainly unsatisfactory, the very strong correlation (96%) between the two solutions can be regarded as an encouranging result for this first test.

Further analysis of the error sources with different gridding and cap amplitude, and specially with denser and wider data sets are to be undertaken before the method could be decleared as viable for local geoid determination, as the authors believe to be worth.

REFERENCES

Arabelos D. (1988): Private comunications.

Holota P. (1982): A Contribution to Mixed Boundary Problems, Int. Symp. Current Methods in Geodesy and Astronomy, Leningrad.

Sacerdote F., Sansò F. (1983): A Contribution to the Analysis of the Altimetry-Gravimetry Problem, Bull. Geod., n. 57, pp. 257-272.

Sacerdote F., Sansò F. (1987): New Developments of Boundary Value Problems in Physical Geodesy, Proc. of the IAG Symposia, IUGG XIX General Assembly, Vancouver, Tome II, pp. 369-390.

Sansò F. (1981): Recent Advances in the Theory of the Geodetic Boundary Value Problem, Rev. Geoph. Space Physics, n. 19, pp. 437-449.

Sansò F., Stock B. (1985): A Numerical Experiment in the Altimetry-Gravimetry Problem II, Manuscripta Geodaetica, n. 10, pp. 23-31.

Schrama E.J.O. (1989): The Role of Orbit Error in Processing of Satellite Altimeter Data, Netherlands Geodetic Commission, Pubblications on Geodesy, n. 33.

Tscherning C.C. (1984): Local Approximation of the Gravity Potential Least-Squares Collocation, Proc. of the Beijing International Summer School on "Local Gravity Field Approximation", Beijing, pp. 277-362.

Wentzel H.G. (1985): Hochauflöesende Kugelkunktionsmodelle für das Gravitationspotential der Erde, Wiss. Arb. der Fachrichtung Vermessungswesen de Universitat Hannover, Hannover, n. 137.

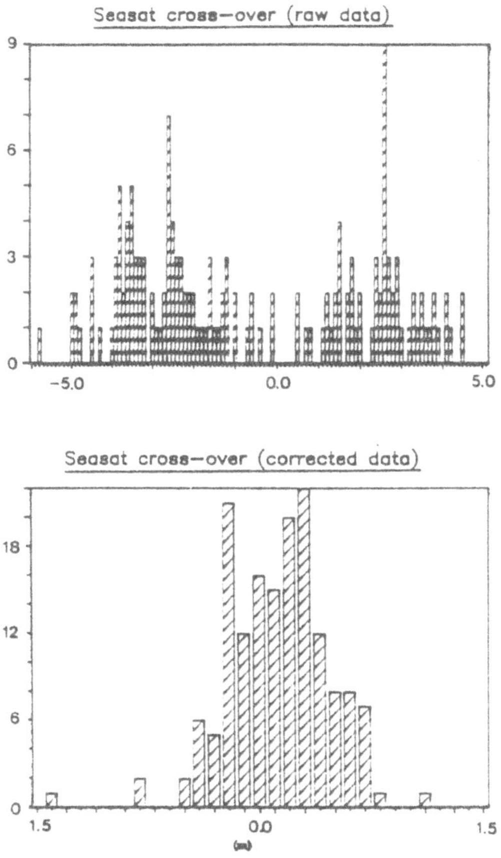

Fig. 1 - SEASAT cross-over.

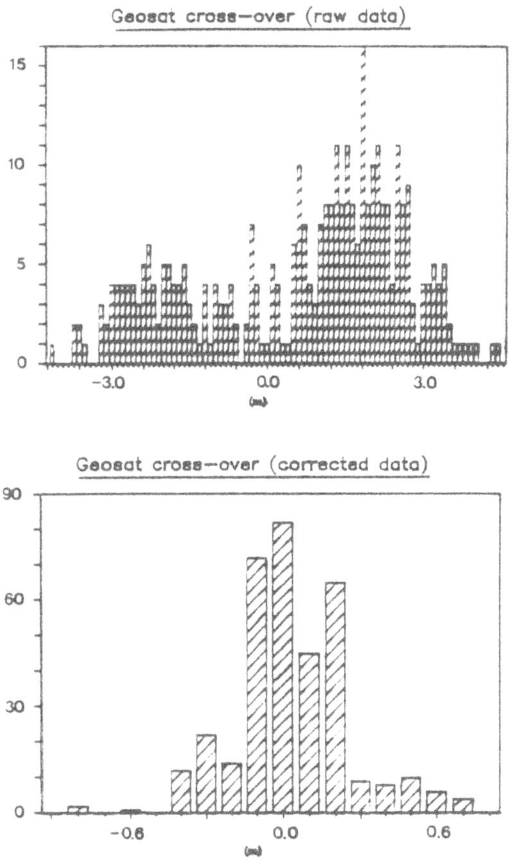

Fig. 2 - GEOSAT cross-over.

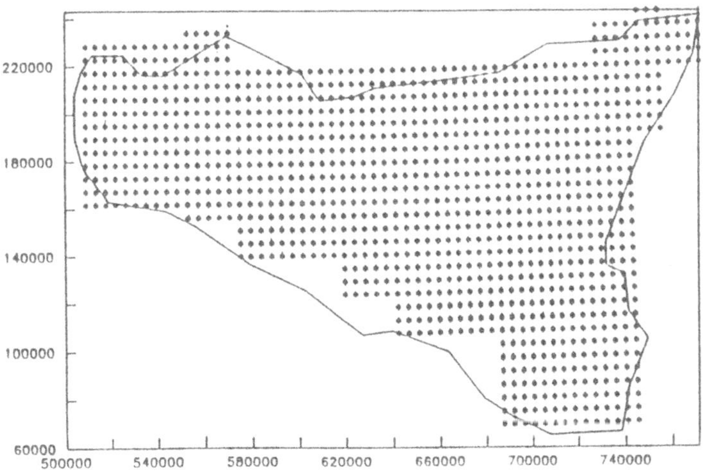

Fig. 3 - Prediction grid.

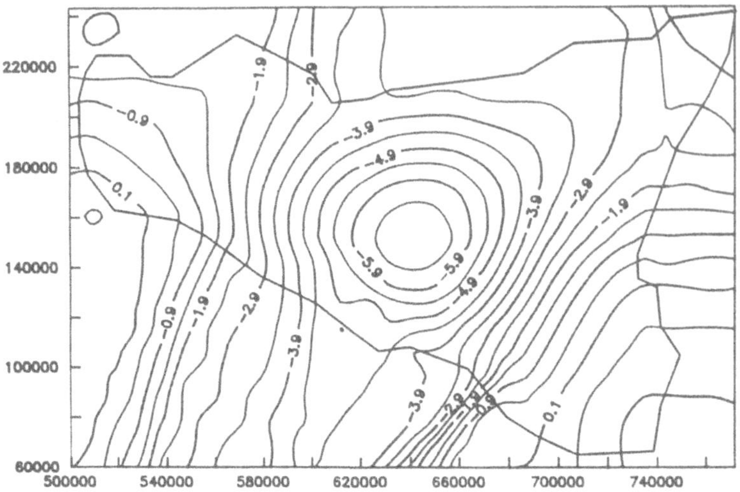

Fig. 4 - Residual geoid (m) with altimetry-gravimetry problem solution.

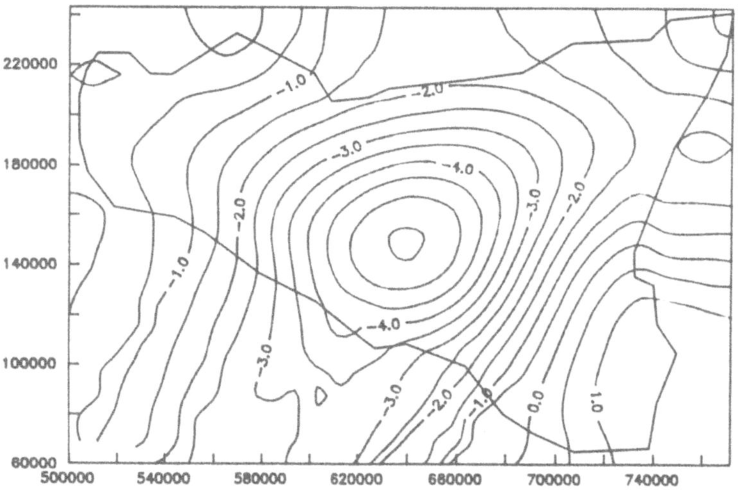

Fig. 5 - Residual geoid (m) with collocation solution.

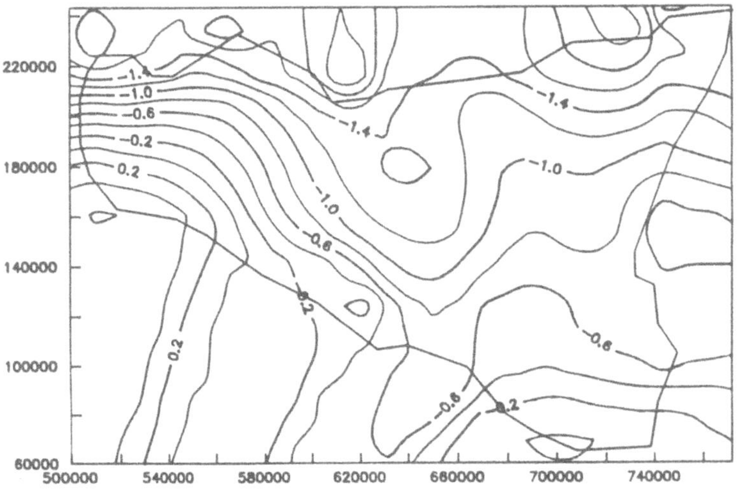

Fig. 6 - Difference between residual geoids (m).

Sea Surface Height Determination in the Mediterranean Sea by Local Adjustment of GEOSAT Altimeter Data

D. ARABELOS, I.N. TZIAVOS
Department of Geodesy and Surveying
University of Thessaloniki
540 06 Thessaloniki, Greece

and

RENE FORSBERG
Geodaetisk Institut
DK 2920 Charlottenlund
Denmark

ABSTRACT

The pre-processing and the crossover adjustment of GEOSAT altimeter data are carried out in order to obtain a precise approximation of the gravity field in the Mediterranean sea. Altimeter data from the U.S. National Oceanographic Data Center (NODC) have been used covering a time period of about one year . With the rejection of outliers and sub-satellite points whose standard deviation (s.d.) exceeds ±0.09 m, the s.d. of the crossover differences was ±3.5 m. After applying a fit and crossover adjustment model, the s.d. of the crossover differences decreased to ±0.16 m. In order to test the efficiency of the sea surface heights performed, a number of predictions was carried out using as data: (a) adjusted satellite altimetry and (b) sea-gravimetry. Least squares collocation was then used to predict subsets of the data types (a) and (b) from subsets of the same data types. The results given here are in terms of s.d. of observed minus predicted values. Using adjusted altimeter data for gravity prediction a s.d. of ±9.36 mgal was found. When gravity values spaced 10' apart were used to predict, (1) gravity values in a grid with the same mesh width shifted 5' and (2) altimeter observations, a s.d. of ±5.08 mgal and ±0.18m, respectively, were found. A combination of gravity and altimeter data made the first s.d. of ±5.08 mgal decrease to ±4.34 mgal showing the impact of combining the two data types. In all the above prediction computations the use of topographic-isostatic effects has been taken into account.

INTRODUCTION

Satellite altimetric data of GEOS3 and SEASAT from the 1970's have contributed to the improvement of our knowledge of the geoid and the gravity field of the earth. Experience with GEOS3 and SEASAT has demonstrated the enormous potential of altimetry and its efficiency in approximating precise marine geoids in open oceans and in limited sea zones surrounded by land areas, such as the Mediterranean sea. The GEOSAT mission was primarily aimed at improving the marine gravity field in local and global applications. In this paper, data from the GEOSAT mission called Exact Repeat Mission (ERM), was used.

The test area is a very interesting but complex zone from a geodynamic and tectonic point of view, presenting an extremely rough and irregular gravity field. Tidal information for the area is not readily available, and the existing SEASAT and GEOSAT geophysical data records (GDR's) are not corrected for tidal effects (see, e.g., Arabelos and Tscherning, 1988). In this paper the tides are considered as constant along the satellite tracks due to the low amplitudes (generally, the tidal amplitudes are lower than 0.20 m in the region). Additionally, the use of ERM data covering an extended time period (about one year in our case) allowed us to neglect the influence of various oceanographic phenomena (e.g., currents, water circulation) on the finally adjusted altimetric data.

The main goal of this paper is to obtain more accurate sea surface heights for the Mediterranean sea taking advantage of the most precise satellite altimetry existing at the time being, namely the ERM of GEOSAT. A good knowledge of the geoid for the test area is necessary to study the oceanographic effects (see, e.g., Bernard et al., 1983).

It is noted that there is a real danger in performing local adjustments involving very short orbit arcs, as those present in the Mediterranean (Cruz and Rapp, 1982). The orbit errors of neighboring arcs heading in the same direction are correlated, and probably the adjusted parameters could be incorrectly obtained. The customary way to remove some of the phenomena which occur in the adjusment of very sort segments of tracks is to keep one or more tracks fixed. This constraint adjustment would allow the arcs to freely adjust among themselves, but the decision on which tracks to be fixed is not an easy one to make (Tai, 1988).

Cruz and Rapp (1982), taking into account the peculiarities of the area under study, the irregularities of the gravity field, and the circumstances already mentioned, have adjusted the SEASAT data for the Mediterranean using only bias parameters.

In this study, an adjustment procedure incorporating crossover constraints was used to fit the sea surface heights (geoid undulations) implied by the altimeter data to the geoid undulations implied by the OSU81 (Rapp, 1981) geopotential coefficients . The error model used for the fit was a two -parameter model (bias and tilting), which was designed to remove altimeter bias and orbit error. To enforce the crossover constraints (or observation equations), an appropriate weight factor was computed from the standard deviation of each observation following a usual regression technique. The above adjustment model with slight modifications has been used by Rapp (1977,1979), Vermeer (1983), Fukuda et al. (1988), and is described in detail by Rowlands (1981).

GEOSAT DATA

The GEOSAT was launched in March 1985. The acronym GEOSAT (GEOdetic SATellite) means that the satellite's primary purpose was the improvement of the gravity field. The first 18 months of observations were classified and this part of GEOSAT's mission called the geodetic mission. However, the geodetic mission ended September 30, 1986. After a series of maneuvres to alter the orbit to produce sea surface profiles within a few kilometers of previeously released SEASAT data tracks, a new generation of data sets was obtained. This new orbit has a repeat period of 17.05 days, and the mission is now referred to as the GEOSAT ERM. The ERM officially became operational November 8, 1986. Its groundtracks are designed to repeat within a 1 km band after 244 revolutions, i.e., 17.05 days. The collinearity of the groundtrack is maintained to a few hundred meters per month. The resolution at the equator is approximately 150 km, so an excellent sample of the ocean is obtained. All the ERM GEOSAT data are unclassified. All this information described here is based on a technical report of NOAA (Cheney et al., 1987).

From the ERM unclassified operational period a number of 90990 sub-satellite points has been used covering the period from February 2, 1987 to December 19, 1987.

In a pre-processing stage, 110 outliers were rejected. These sub-satellite points were detected at the edges of the arcs, i.e., close to land areas, according to their unexpected negative value. Then, the sub-satellite points were coded according to a number describing the revolution number multiplied by 10000 plus the sequence number of the point within the revolution. The revolution number was evaluated as follows:

$$\text{revolution number} = t \times 244 \; / \; 24 \; / \; 17.05 \, ,$$

where t is the current UTC time, in hrs, with respect to the beginning of 1985.

After the rejection of outliers, a number of sub-satellite points with an internal s.d. exceeding ±0.09 m was removed. This was done in order to obtain sea surface heights with an accuracy better than 0.1 m. The number of remaining points was 88670, distributed in 840 arcs. All these arcs were less than 236 seconds (about 1604 km) long, resulting in

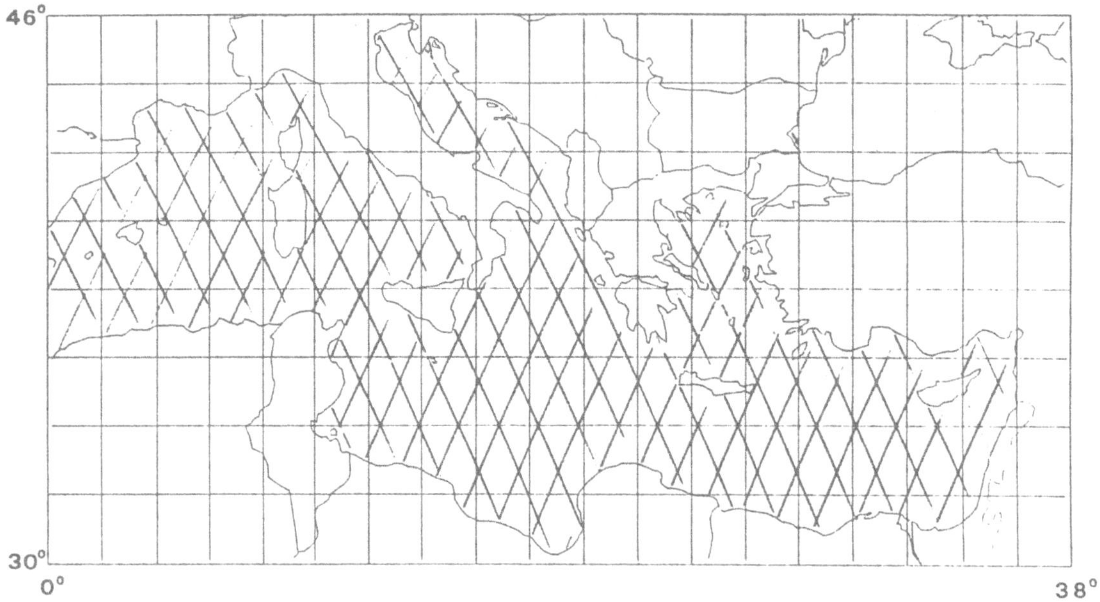

Figure 1. GEOSAT altimetry in Mediterranean sea.

the pattern of Figure 1. The above unadjusted data heights were compared with corresponding geoidal heights derived from the OSU81 geopotential model. The result from this comparison (GEOSAT-OSU81) indicated an offset of the order of 1.6 m, a s.d. of differences of ±3.3 m, a minimum value of differences of -11.4 m and a maximum value of + 16.2 m.

CROSSOVER ANALYSIS

For the processing of the above described data, a crossover analysis was carried out. In the intersection of ascending and descending arcs the height differences were determined in the sense: "ascending-descending". An error estimation for each crossover was also found based on the standard deviation of the neighboring points and the distances between them (see, e.g., Knudsen 1987). This error estimation will be used later to give proper weights to the observation equations.

The position of the fixed intersection point was found by linear interpolation between the four neighboring data points. In each intersection point a separation value, in km, was found for each arc with respect to a common origin (the parallel of 37°). According to the above scheme for each crossover point, two separation values were determined which represent the distance of intersection from the origin.

Due to the large number of available data, we adopted crossovers where the distance between two neighboring points of each of the two crossing arcs was less than 16.75 km (=2.5 times the mean separation between two successive sub-satellite points). Finally, 20427 crossovers were found, with a s.d. crossover discrepancy ±3.3 m before the adjustment, and a mean value of differences -2.13 m. This means that the descending arcs had a large offset (too high) of the order of 2 m with respect to the ascending arcs.

ADJUSTMENT MODEL AND RESULTS

In our local adjustments we adjusted the area within the limits $30° \leq \varphi \leq 50°$, $0° \leq \lambda \leq 37°$. This area included the Mediterranean sea. In this study the fit and crossover adjustment model was adopted which takes into consideration the existing long wavelength information of the gravity field (e.g, a spherical harmonics model). A similar procedure has been utilized by Rapp (1977). Practically, this model can be split up into two parts. The first part is a fitting of sea surface heights to a reference geoid along each orbit arc, and the second part is a minimization of the discrepancies at crossover points of ascending and descending arcs.

The observation equation of the model can be written as:

$$h_i - h_j = d_{ij} = a_i - a_j = a_i - a_j + b_i s_i - b_j s_j \tag{1}$$

where:

a_i, a_j are the i-th and j-th arc's bias parameters,

b_i, b_j are the corresponding tilting parameters with respect to a common origin

$d_{ij} = h_i - h_j$ is the difference between the observed sea surface heights corresponding to i,j arcs at the crossover point , and

s_i, s_j are the distances, in km of, the crossover point along each track with respect to a common origin which is valid for all tracks.

It is possible to compute approximate values a^o, b^o for the parameters a, b by comparing GEOSAT sea surface heights with OSU81 geoidal heights according to the model

$$h - H = a^o + b^o s \tag{2}$$

where H is the geoidal height derived from the geopotential model OSU81. Introducing the approximate values a^0, b^0 into equation (1) the following equation results:

$$d_{ij} - (a_i^0 + b_i^0 s_i - a_j^0 - b_j^0 s_j) = \delta a_i - \delta a_j + \delta b_i s_i - \delta b_j s_j + v_{ij} \tag{3}$$

where: $\delta a = a - a^0$, $\delta b = b - b^0$.

It also holds:

$$b_{ij} = [\,1 \quad -1 \quad s_i \quad -s_j\,] \begin{vmatrix} \delta a_i \\ \delta a_j \\ \delta b_i \\ \delta b_j \end{vmatrix} + v_{ij} \tag{4}$$

where

$$b_{ij} = d_{ij} - (a_i^0 + b_i^0 s_i - a_j^0 - b_j^0 s_j) \tag{5}$$

It is noted that the number of equations (3) is equal to the number of crossover differences (m), the number of unknowns is equal to the number of involved arcs (n) and v (mx1) is the residual vector.
For the total number of observations, the system of observation equations can be written in the following convenient matrix form:

$$\mathbf{b} = \mathbf{A}\mathbf{x} + \mathbf{v} \tag{6}$$

where:

A is the (mxn) matrix of the coefficients a_i, a_j, b_i, b_j which take the value 1 and -1
x is the (nx1) vector of the a_i, a_j, b_i, b_j values
b is the (mx1) vector of the crossover differences

The corresponding minimum condition is

$$\mathbf{v}^T\mathbf{P}\mathbf{v} + \mathbf{x}^T\mathbf{P}_\mathbf{x}\,\mathbf{x} = \min \tag{7}$$

where $\mathbf{P_x} = \mathbf{C_x^{-1}}$ is the weight matrix of parameters of the vector $\mathbf{x}$, and $\mathbf{C_x}$ is the covariance matrix of quantities a^o, b^o that resulted from the comparison between GEOSAT and OSU81. By using apriori variances on the parameters $\mathbf{x}$ the final adjustment solution is thus forced to yield the same overall bias and tilt in the sea surface height as implied by the OSU81 geoid. Condition (6) leads to the following system of generalized inverse normal equations written in matrix form (Dermanis 1987) as:

$$\hat{\mathbf{x}} = (\mathbf{A}^T\mathbf{P}\mathbf{A} + \mathbf{P_x})^{-1}\,(\mathbf{A}^T\mathbf{P}\mathbf{b}) \tag{8}$$

The approximate values a_i, b_i derived from the comparison of sea surface heights of i-th orbit with OSU81 geoidal heights using equation (2) can be computed from the solution of the following system:

$$\begin{vmatrix} \sum_{k=1}^{m} p_k & \sum_{k=1}^{m} s_k p_k \\ \sum_{k=1}^{m} s_k p_k & \sum_{k=1}^{m} s_k^2 p_k \end{vmatrix} \begin{vmatrix} a_i^o \\ b_i^o \end{vmatrix} = \begin{vmatrix} \sum_{k}^{m}(h_k - H_k) \\ \sum_{k}^{m} s_k(h_k - H_k)p_k \end{vmatrix} \tag{9}$$

where $k=1,2,3,\ldots\ldots,m$ is the number of sub-satellite points along i-th orbit, s is the distance of each point from a common origin and p_k is the appropriate weight that introduces each sub-satellite point into the specific observation equation. From the above solution, the elements $\sigma^2_{a_i}$, $\sigma^2_{b_i}$, $\sigma_{a_i b_i}$ of matrix $\mathbf{C_x}$ can be computed by the formula:

$$\begin{vmatrix} \sigma^2_{a_i} & \sigma_{a_i b_i} \\ \sigma_{a_i b_i} & \sigma^2_{b_i} \end{vmatrix} = \begin{vmatrix} \sum p_k & \sum s_k p_k \\ \sum s_k p_k & \sum s_k^2 p_k \end{vmatrix}^{-1} \tag{10}$$

The cross-covariances $\sigma_{a_i a_j}$, $\sigma_{a_i b_j}$, $\sigma_{b_i b_j}$, $\sigma_{b_i a_j}$ are considered equal to zero.

In the first step of the previously theoretically developed model a linear fit of the sea surface heights to the OSU81 geopotential model was thus made. Arcs including more than 30 sub-satellite points have been used in these computations . The parallel of 37° was utilized as a common origin to measure the distances mentioned in equations (1), (2), (3). Due to computer storage limitations, only 41897 sea surface heights were actually used in this method. The statistics of the discrepancies between these original sea surface heights, and the corresponding OSU81 heights, are summarized in Table 1. In the same table the results obtained from the regression analysis are presented. It is obvious that the descending arcs are systematically higher than the ascending arcs, and present a slightly larger tilting coefficient. The correction of each orbit arc was made according to the coefficients a, b for each arc, and the same comparison with OSU81 heights was repeated. These values are also given in Table 1.

Table 1. Statistical results from the 1st step of the adjustment

	GEOSAT-OSU81 h	H	REGRESSION Ascending arcs		Descending arcs	
	Before	After Regression	a	b	a	b
Nr.	41897	41897	216	216	203	203
mean value	1.27 m	-0.005m	0.28m	0.0029m/km	3.41	0.0035m/km
s.d.	±2.60 m	±0.71m	±2.46m	±0.0028m/km	±3.71	±0.0043m/km

In the second step of the model, the crossover adjustment of the above corrected orbit arcs was carried out. The results of the crossover analysis are given in Table 2. The results of the crossover adjustment are given in the same Table as well as the statistical results from the geoidal height comparison (GEOSAT - OSU81) after this final adjustment.

Table 2. Statistical results from the 2nd step of the adjustment

	Crossovers Before Adjustment	Crossovers After Adjustment	GEOSAT - OSU81(h - H) After Adjustment
Nr	5517	5517	41897
mean [m]	-0.007	-0.002	0.00
s.d. [m]	±0.55	±0.16	±0.76

From the analysis of the results derived from the model used, and taking into account conclusions from recent studies (see, e.g., Arabelos and Tziavos 1989), it was found that an adjustment based only on crossover information presents disadvantages in restricted areas such as the Mediterranean employing short arcs which in many cases are interrupted by land areas (e.g., islands, peninsulas). Arabelos and Tziavos (1989), comparing Cruz and Rapp's (1982) SEASAT adjustment and a pure crossover adjustment with the OSU81 solution, both referring to the same test area, found that Cruz and Rapp's solution fits better OSU81 with regard to both the mean value as well as the s.d. of the differences. This can be attributed to the fact that in Cruz and Rapp's adjustment three arcs were fixed while in the crossover solution a free adjustment was performed.

Using fit and crossover equations (present study) a very good agreement was obtained with the reference geoid (OSU81). In this case the mean value of the differences was 0.0m and the s.d. was ±0.76 m, while the corresponding quantities using the crossover adjustment model were 1.70 m and ±1.85 m, respectively (Arabelos and Tziavos 1989). Assumming that OSU81 represents succesfully the gravity field in the test area in the lower frequency part of the gravity spectrum (see, e.g., Arabelos and Tscherning, 1988), the fit and crossover adjustment model gives better results than a pure crossover model.

PREDICTION TESTS AND RESULTS

In order to test the compatibility of the adjusted GEOSAT altimeter data with the sea-gravity data a number of prediction tests was carried out.

The area selected for our prediction tests is bounded by [34° ≤ φ ≤ 38°, 16° ≤ λ ≤ 22°] and presents strong geophysical signals. In this area the following data were available (see Fig. 2): (a) 4578 altimeter GEOSAT sea surface heights presenting a mean value equal to 0.34 m and a s.d. equal to ±0.54 m after the subtraction of the contribution of the OSU81 field. (b) 3401 point gravity values spaced 5' apart (obtained from digitized gravity maps) presenting a mean value equal to -1.1 m and a s.d. equal to ±20.83 mgal. From the data (a), (b) the influence of the isostatic effect was removed according to the findings of a previous study for Eastern Mediterranean (Arabelos and Tscherning 1988). After the isostatic reduction, the statistic for the GEOSAT altimetry made: mean value = 5.22 m and s.d. = ±0.94 m, and for the gravity: mean value = 22.55 m, s.d. = ±18.81m.

Two subsets for each kind of the above mentioned data were prepared. The data of the first subset covering the area 34° ≤ φ ≤ 38°, 16° ≤ λ ≤ 22° was used as sample points and the data of the second subset covering the area 35°≤ φ ≤ 37°, 18° ≤ λ ≤ 20° was used as control points.

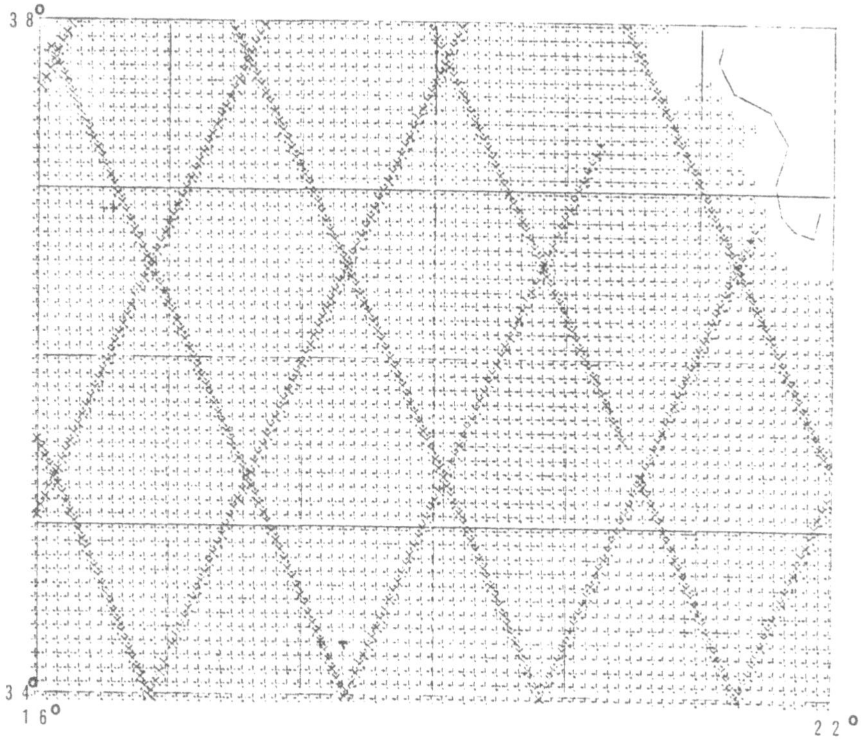

Fig. 2. The available data in the prediction test area: GEOSAT altimeter data tracks (x), and gravity data at 5' spacing (+).

The subset of the gravity sample points was formed using every second value of the 5'x5' original grid. This resulted a 10'x10' grid. The remaining points formed the subset of control points.

From the adjusted GEOSAT data only the points included in one 17-day period were used in our computations. The subset of the sample points was formed taking points as close as possible to the knots of a 10'x10' grid. The remaining points were used as the subset of the control points.

The starting point for the use of least squares collocation in our local application was the computation and analytic modelling of the empirical covariance functions. Firstly the auto- and cross- covariance functions for gravity anomalies and sea surface heights were carried out. Then, the modelling of the empirical values using the second analytical model of Tscherning (1975) was carried out. The following values for the parameters characterized the covariance function were estimated:

variance of point gravity anomalies = 410 $mgal^2$,

$R_E - R_B$ = 3.5 km (R_E=6371 km and R_B is the radius of the so-called Bjerhammar sphere)

Having developed analytic models for the covariance functions the FORTRAN program GEOCOL (Tscherning 1987) was used to predict altimeter heights and gravity anomalies from various combinations of the selected data mentioned previously.

Three prediction tests were carried out using as sample points (a) gravity alone; (b) altimetry alone; (c) gravity and altimetry simultaneously. The results are summarized in Table 3.

Table 3. Prediction results of tests using altimetry, gravity and both in combination

		altimetry obs. - pred.		gravity obs. - pred.	
		(62 points)		(522 points)	
input data	number of points used	mean [m]	s.d. [m]	mean [mgal]	s.d. [mgal]
altimetry	243	0.01	±0.08	6.45	±9.36
gravity	639	1.27	±0.18	6.72	±5.08
combination	882	0.02	±0.08	3.27	±4.34

Note here that the data combination gives the best results when gravity anomalies or altimetric heights were predicted.

The results of this study are comparable with the results obtained by Arabelos and Tscherning (1988), although the altimetric coverage of the test area is poorer than the SEASAT altimeter data coverage in the Eastern Mediterranean (Arabelos and Tscherning 1988).

The results performed in this study show the compatibility of the two data types (gravity, altimetry) and the impact of combining these data types for prediction purposes.

CONCLUDING REMARKS

Satellite altimetry in a closed sea like the Mediterranean must be carefully checked for bias and tilt parameters. Initial difficulties with the adjustment of altimeter data in the test area appear to have been caused by the isolation of the region from arcs in open seas, like the Atlantic and Indian Ocean.

To obtain an improved sea surface, a combined model, namely fit and crossover equations using simultaneously bias and tilt parameters were applied. We slightly modified the classical procedure by introducing proper weights derived by a regression analysis. The s.d. of the crossover differences after the adjustment was 0.16 m.

The prediction tests indicated the compatibility of the adjusted GEOSAT sea surface heights with sea-gravity, as well as that the GEOSAT ERM data can be characterized as the most accurate satellite altimetric mission up to now.

REFERENCES

Arabelos D. and C.C. Tscherning (1988). Gravity field mapping from satellite altimetry, sea-gravimetry and bathymetry in the eastern Mediterranean. *Geophysical Journal*, 92, 195-206.

Arabelos D. and I.N. Tziavos (1989). Sea Surface Heights in the Mediterranean Sea from GEOSAT Altimeter Data. Submitted for publication *J. Geophys. Res.*

Bernard, J., F. Barlier, J.P. Bethoux, M. Souriau (1983). First Seasat altimeter data analysis on the western Mediterranean sea. *J. Geophys. Res.*, 88, (C3), 1581-1588.

Cheney, R.E., B.C. Douglas, R.W. Agreen, L. Miller, D.L. Porter, M.S. Doyle (1987). Geosat altimeter geophysical data record user handbook. *NOAA technical memorandum NOS NGS-46*.

Cruz, J.Y., and R.H. Rapp (1982). Sea surface heights in the Mediterranean area from Seasat altimeter data. *Boll. Geof. Teor. Appl.*, XXIV, (95), 167-174.

Dermanis A. (1987). Adjustment of observations and estimation theory. In Greek.

Fukuda, Y., J. Segawa, H. Watanabe, and A. Iwasita (1988). Geosat altimeter data processing. *Journal of the Geodetic Society of Japan*, 34, (2), 97-108.

Knudsen, P. (1987). Adjustment of satellite altimeter data from crossover differences using covariance relations for the time varying components represented by Gaussian functions. In *Proceedings of the IAG Symposia, XIX General Assembly*, Vancouver, Canada, Aug. 10-22, Vol. II, 617-628.

Rapp, R. H. (1977). Mean gravity anomalies and sea surface heights derived from Geos3 altimeter data. *Dept. of Geod. Sci. Surv. Rep.* 325, Ohio State Univ., Columbus.

Rapp, R. H. (1979). Geos3 data processing for the recovery of geoid undulation and gravity anomalies. *J. Geophys. Res.*, 84, (B8), 3784-3792.

Rapp, R. H. (1981). The earth's gravity field to degree and order 180 using Seasat altimeter data, terrestrial gravity data, and other data. *Dept. of Geod. Sci. Surv. Rep.* 322, Ohio State Univ., Columbus.

Rowlands, D. (1981). The adjustment of Seasat altimeter data on a global basis for geoid and sea surface height determinations. *Dept. of Geod. Sci. Surv. Rep.* 325, Ohio State Univ., Columbus.

Tai, C.-K. (1988). Geoidal crossover analysis in the Tropical Pacific. 1. Constrained sinusoidal adjustment. *J. Geophys. Res.*, 93, (C9), 1 0621-10629.

Tscherning, C.C. (1975). Covariance expressions for second and lower order derivatives of the anomalous potential. *Dept. of Geod. Sci. Surv. Rep.* 225, Ohio State Univ., Columbus.

Vermeer, M. (1983). A new Seasat altimeter geoid for the Baltic. *Finnish Geodetic Institute, Rep.* 83, (4), 1-10.

THE MEAN SEA LEVEL OF THE GULF STREAM ESTIMATED FROM SATELLITE ALTIMETRIC AND INFRARED DATA.

Victor Zlotnicki

**Jet Propulsion Laboratory,
California Institute of Technology
Pasadena, CA. 91109. USA.**

INTRODUCTION

Altimeters aboard satellites such as Seasat and Geosat, flying around 700 km above sea level, measure sea level at a rate of one sample averaged over approximately 6.7 km every second. Such sampling has yielded unprecented coverage in deep water where tide gages cannot be placed and in ocean regions where weather discourages ship presence, at a rate 1400 times faster than a ship carrying a gravimeter (although the satellite has somewhat less alongtrack resolution for gravity than a ship does). These data have been applied to a wide variety of geophysical and oceanographic applications (e.g., Douglas et al., 1987).

The signal measured by the altimeter, after a variety of instrumental, atmospheric, and orbital errors are reduced to a tolerable level, is the sum of the geoid undulations (N, with 25 m RMS), time averaged oceanographic signals (ζ, with 0.3 m RMS) and time-varying oceanographic signals. Geophysical studies make no attempt to remove the time-averaged oceanographic signal, since for those applications it is a tolerable error. Most oceanographic studies have concentrated on the time-varying signal, whose energy on the observed time scales is much larger than the energy in time changes of the geoid.

Separating ζ and N in altimetric mean sea surfaces, however, has remained an interesting endeavour practiced by a few, both for its challenge and because the horizontal gradients of ζ measure the average surface velocity of the large scale ocean currents that carry heat, nutrients, and tracers. Such studies fall naturally into two ranges of length scales, separated around spherical harmonic degree 10.

Non-dynamic long wavelength separations have simply used a mean sea surface from altimetry, a long wavelength geoid estimate from satellite orbit perturbation data, and clever filtering of high degree spherical harmonics to distinguish the two components (Tai, 1984, 1988; Engelis, 1985). The estimate of ζ thus obtained is then compared to a similarly filtered estimate of ζ obtained from classical oceanographic measurements of temperature and salinity of seawater (see Wunsch and Gaposhkin, 1980, for an excellent review).

Dynamic long wavelength separations have relied on the different effect that a bump in the sea surface (observed with an altimeter) has on a satellite's orbit (observed with lasers, range rate, and altimetry) depending on whether the bump is a gravitational feature (which affects the orbit, if its scale is sufficiently large) or an oceanographic one (with negligible effect on the orbit at all scales). The theory was demonstrated by Wagner (1986) and Engelis (1987), and implemented by Tapley et al. (1988) and by Marsh et al. (1989).

Short wavelength separations have used ship gravity data to distinguish between N and ζ. Cheney and Marsh (1981) substracted a short wavelength geoid from altimetry. Zlotnicki (1984) used a geoid and a mean sea surface in a simple collocation estimate, and concluded that the geoidal omission error (insufficient gravity coverage) exceeded ocean signals almost everywhere in the N. Atlantic. Zlotnicki and Marsh (1989) avoided computing a geoid, subtracting instead altimetric gravity distrubances from ship gravity anomalies.

Oceanographic data used to estimate ζ has been used up to now mostly for comparison. The reason, discussed by Wunsch and Gaposhkin (1980), is that the ship sampling of temperature and salinity causes an inevitable aliasing of energetic time varying oceanic motions into the estimate of the time averaged ζ. However, Roemmich and Wunsch (1982) showed that under well sampled conditions, oceanographic data can determine ζ to better than 10 cm.

The work described below is an attempt at combining a) satellite infrared data to determine the positions of a strong meandering current, b) a simple model of the velocity and sea level signal of the current based on hydrographic and water velocity data, and c) satellite altimetric data both to calibrate and to evaluate the model, to d) estimate the time-averaged sea level signal ζ of the current, yielding a correction to altimetry, valid in a small geographic range, in order to retrieve N.

GULF STREAM SEA LEVEL AND GEOSTROPHIC VELOCITY MODEL

The Gulf Stream is in the same place, *on average* and within a few hundred km, as it was when Franklin and Folger first mapped it 218 years ago (see Richardson, 1980), and as such it is a good example of the concept 'time averaged general circulation of the oceans'. However, the current meanders, sheds eddies, and appears to change intensity seasonally, so that an observer at a fixed latitude-longitude point within the current will see sea level change rapidly in time because of the meanders and other current changes (Fofonoff, 1981; Flierl and Robinson, 1984). While these time changes cannot be described with geostrophy, which is a steady state simplification, the downstream velocity component of the Gulf Stream, v, is in geostrophic balance with the cross stream pressure gradient, to an excellent approximation (e.g., Joyce et al., 1989), so we can write

$$v = (-g/f)\,(\partial h/\partial x) \qquad (1)$$

where $g \sim 980$ cm s^{-2}, $f = (4\pi/86400)\sin(\text{latitude})$ s^{-1}, h is the height of sea level relative to the geoid, and x is the crosstream cartesian coordinate. Equation (1) only holds for the components of v and h with wavelengths larger than about $2.\pi.30$ km (the first baroclinic Rossby radius in midlatitudes) and periods longer than about 1 day.

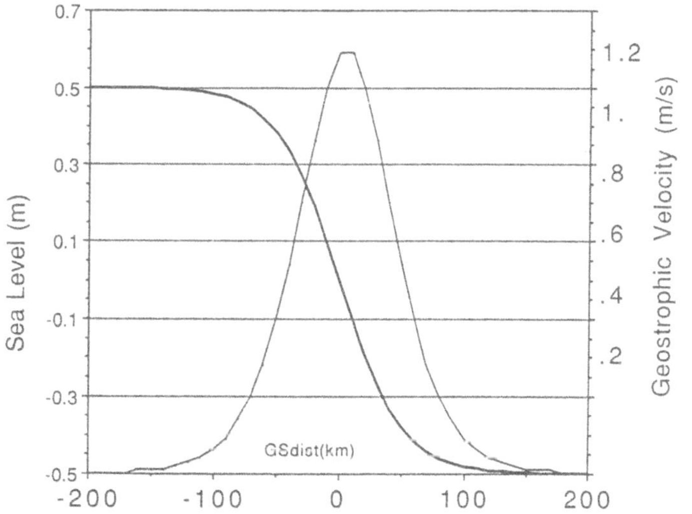

Figure 1. Gulf Stream sea level and surface velocity as a function of cross stream distance.

Several studies show that the density profiles across the Gulf Stream, and thus the sea level profile, can be approximated by a step with rounded corners , higher to the south and east of the current (e.g., Joyce et al (1986); Worthington (1976) figure 3). We used H tanh (x/L), with H= 50 cm, L=50km (figure 1), with a corresponding velocity profile $V = \mathrm{sech}^2(x/L)$, V~ 1.2 m s^{-1}. The surface geostrophic velocity being the corresponding derivative from equation (1), is a hyperbolic secant squared. Kelly and Gille (1989) used a gaussian velocity model (sea level is an errfc) to demonstrate how time changes in the Gulf Stream suface velocities could be computed from Geosat altimetry. Vazquez et al (1989) showd that a tanh approximation describes the annual cycle of sea level variability around the Gulf Stream. Tai (1989, pers. comm.) has used a gaussian velocity model with a one step nonlinear inversion to estimate surface velocity of the Gulf Stream from altimetry.

ALTIMETRIC AND INFRARED DATA.

The U.S. National Oceanic and Atmospheric Administration's (NOAA) National Climatic Data Center (NCDC) releases weekly charts of the position of the North Wall of the Gulf Stream and of selected eddies in the area, based on sea surface temperature measured from the Advanced Very High Resolution Radiometer (AVHRR) on board the NOAA-10, NOAA-11 satellites. The Gulf Stream edges were digitized from these charts (see Vazquez et al., 1989, for details), and the set of all such edges for 1987 is shown in figure 2.

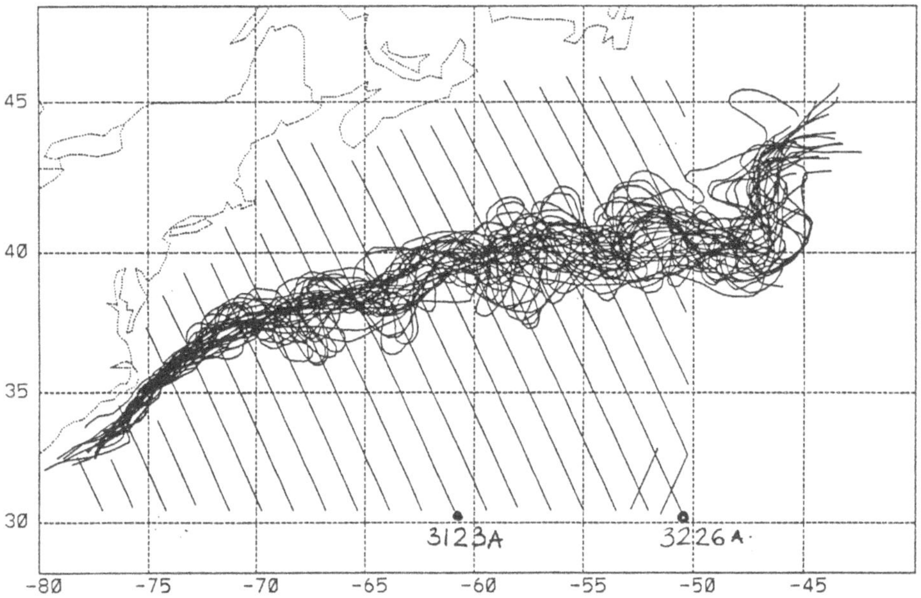

Figure 2. Weekly positions of the North Wall of the Gulf Stream from NOAA AVHRR data, and ground tracks of Geosat altimetry. The descending tracks (NW-SE) are mostly missing because of the great data loss off the continent.

Altimetric data from Geosat, whose tracks are also plotted in figure 2, were used for this study. The data processing steps, atmospheric and orbital corrections applied are described in Vazquez et al. (1989) and Zlotnicki et al. (1989), hence they will not be repeated here. The resulting observations, $h(s(\phi,\lambda),t)$, are sea level heights h at time t, along a track $s(\phi,\lambda)$, from which the following components were removed:

a 1 year sea level average (to get around the lack of an accurate geoid)
a 2500 km alongtrack quadratic (to reduce orbit error).
ocean tides (accurate only in deep water)
the inverse barometer part of the response of sea level to atmospheric pressure
atmospheric corrections to the altimeter pulse propagation.

COMPARING THE DATA AND THE MODEL

Along each ground track $s(\phi,\lambda)$, spaced about 128 km at this latitude, we had data $h(s(\phi,\lambda),t)$ at times t_k, such that t_k-t_{k-1}=17.05 days, or 22 repeats for the year 1987 (because Geosat crosses the Gulf Stream in less than 30 seconds, an interval negligibly small with respect to the observed current's time scale, t_k refers to the time of each repeat). The sea level model in figure 1 yielded a simulated sea level height residual $h'(s(\phi,\lambda),t)$, relative to a one year average:

$$h'(s(\phi,\lambda),t_k) =$$

$$H \tanh((s-s_0(t_k))\cos(\beta(t_k))/L) - \sum_{k'=1}^{22} H \tanh((s-s_0(t_{k'}))\cos(\beta(t_{k'}))/L) \qquad (2)$$

at the times t and positions $s(\phi,\lambda)$ of the Geosat observations. The Gulf Stream position s_0 along track $s(\phi,\lambda)$, and the angles β between the current edge and $s(\phi,\lambda)$ were given by the AVHRR edges, interpolated from the weekly values to Geosat times. The parameters H=50 cm, L=50 km were kept constant. The sum in (2) is performed over all repeats during 1987. We chose to treat pronounced meanders causing multiple crossings (see figure 2) as detached eddies that were eliminated from the calculation, although better options are available.

Figure 3 shows a set of modelled residuals, mostly to familiarize the reader with the signal given by the current once the annual mean is removed. The model is a good approximation to reality, as illustrated by figure 4a, which shows that the root mean squared sea level residuals (averaging across t_k, at fixed latitude-longitude) along two particular tracks obtained through equation (2) fit the corresponding values from Geosat in the 200 km band where the current meanders. Figure 4b illustrates the error introduced by selectively removing multiple crossings. Both figures 4 also show a slight displacement to the north of the modelled rms heights relative to the observed rms height. This displacement, about 20 km, results from the displacement between the surface expression of the North wall, observed with AVHRR, and the vertically (from sea surface to 'level of no motion') integrated density contrast which determines the sea level signal. Using all tracks shown in figure 1, we found that the percentage of variance in sea level explained by the model, within 100 km of the mean current position ranges between 70% and 100% at longitudes between 287º and 309º, but falls to 50% west of 286º, and exceeds 100% east of 310º.

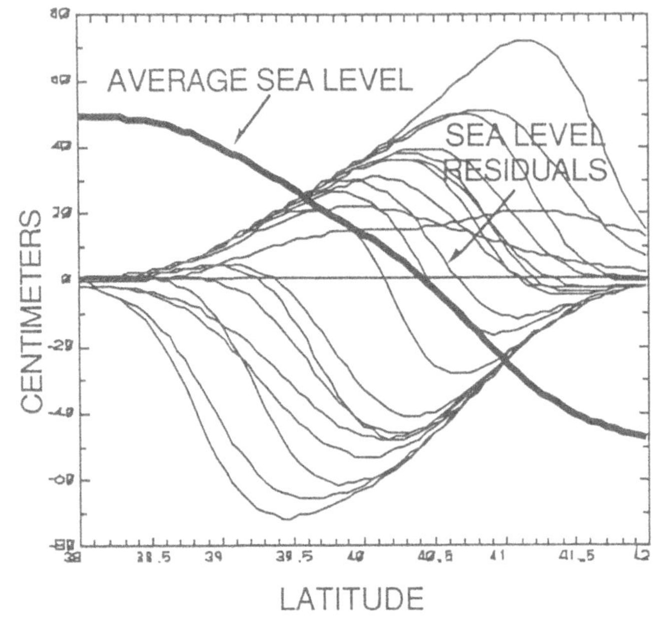

Figure 3. Simulated residuals h'(s(ϕ,λ),t) from the yearly mean along the ascending track that crosses the Equator at 322.6º (+/- 1 km). Notice that current positions south of the mean yield a sea level residual that looks like a cold eddy, current positions to the north of the mean position resemble a warm eddy, and only current positions within 20 km of the mean position cause a step in the sea level residual, but with greatly reduced amplitude.

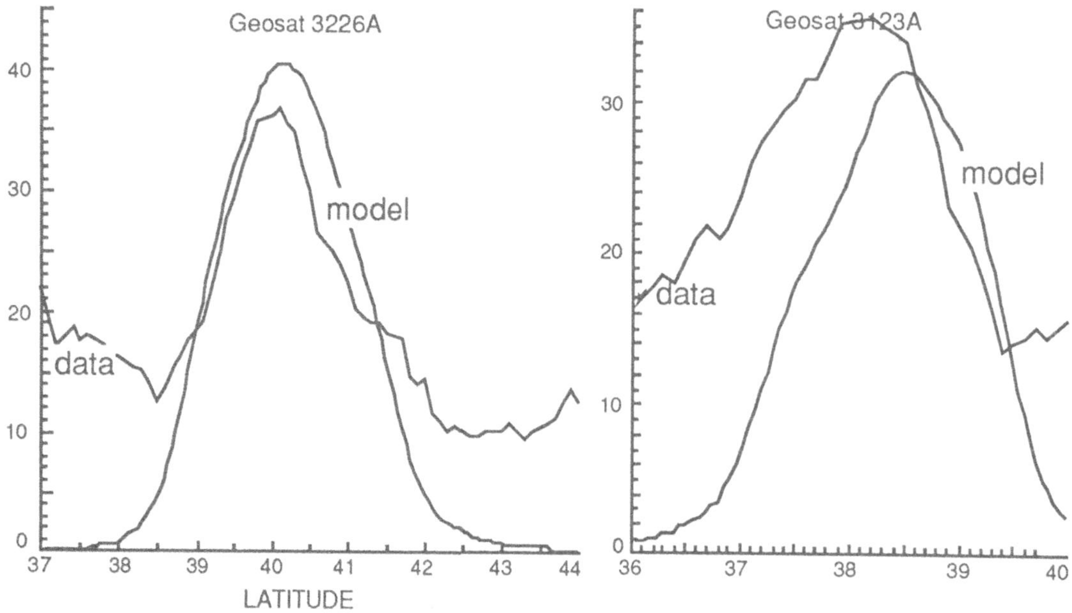

Figure 4. Root mean squared sea level residual (averaged over 1 year, at fixed lat-lon) for two Geosat tracks, from both the altimeter data and the tanh model with AVHRR positions.

The second term in (2), the yearly average sea level above the geoid, cannot be recovered directly from the altimetric data due to the lack of accurate gravimetric geoid estimates, but can be estimated by this modelling approach. The set of such yearly averages from a set of 24 ascending tracks crossing the Gulf Stream between longitudes 285º and 310º were gridded, using a successive approximation mapping scheme (described in Vazquez et al., 1989) yielding the correction map in figure 5. This map is only valid within the region of the -40 cm, + 40 cm contours, and undefined elsewhere, as is the model used to estimate it.

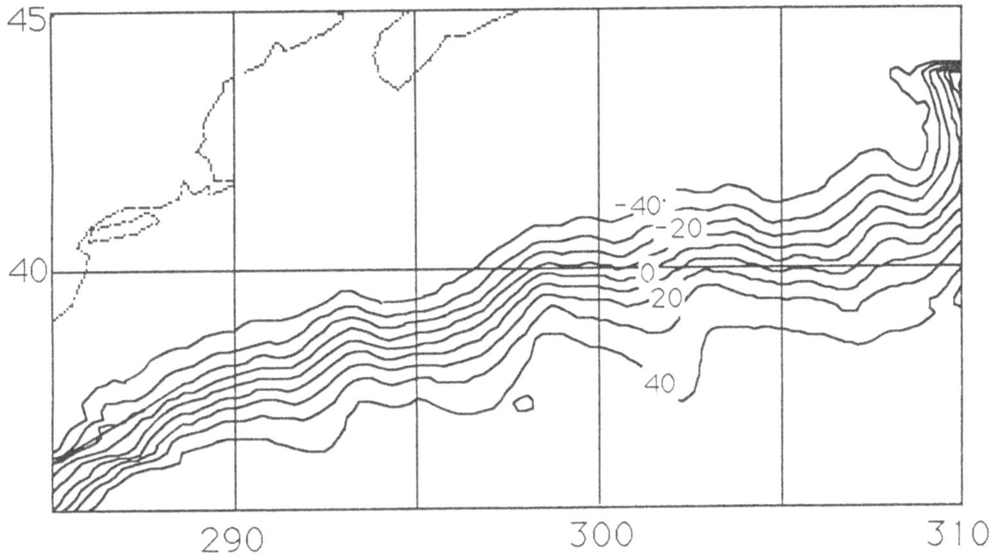

Figure 5. Correction to an altimetric geoid estimate for this region obtained by averaging Geosat data for 1987, based on the Gulf Stream model, AVHRR and Geosat data.

CONCLUSIONS AND DISCUSSION.

It was shown above and in Vazquez et al. (1989) that most of the sea level variability within 100 to 200 km of the Gulf Stream's mean position is due to shifting position of the meandering current, with less than 10% of the signal caused by changes in the current intensity. A simple tanh(x) model of the sea level signal of the Gulf Stream reproduces the statistics of sea level changes observed with altimetry, if the position of the current is defined from satellite infrared measurements of sea surface temperature.

The simple model also yields, as a byproduct, a correction of order 1 m to an *altimetric* geoid estimate, computed in the usual way by averaging altimetric sea level (or slopes) over 1 year or longer in order to reduce altimetric noise and time-varying ocean signals. This approach does not suffer from the aliasing errors inherent in all 'historical average' dynamic topography estimates, whereby ship cruises sampling the current in different seasons of different years in different places are used as if they were simultaneous (Wunsch and Gaposhkin, 1980). This is a relatively small problem at wavelengths *much* longer than $2.\pi.30$km (in midlatitudes), but a strong problem at wavelengths close to the critical value.

This approach would benefit from several improvements. There is no reason to fix the amplitude H and width L of the current. Kelly and Gille (1989) demostrated how to estimate these and other parameters in a several step optimization also using AVHRR data, while Tai (1989, pers. comm.) performed a single step nonlinear optimization (with more data than Kelly and Gille) to estimate the current parameters, but did not use AVHRR data for the initial values. Assuming that those calculations are able to reproduce the observed variability, and that the parameter estimation is robust relative to small changes in the initial values, then they always yield as a byproduct the yearly average sea level above the geoid.

Approaches such as these will continue to improve the short wavelength geoid estimates from altimetry; simultaneous orbit, geoid and dynamic topography estimates from altimetric satellite tracking will continue to improve the long wavelength components until dedicated gravitational missions, such as Aristoteles, are finally launched.

Acknowledgements Lee Fu suggested the tanh profile. Jorge Vazquez digitized the AVHRR edges. This work was performed by the Jet Propulsion Laboratory under contract with the National Aeronautics and Space Administration.

REFERENCES

Cheney, R.E. and J.G.Marsh: , SEASAT altimeter observations of dynamic topography in the Gulf Stream region, J. Geophys. Res., 86:473-483. , 1981

Douglas B.C., McAdoo, D.C. and R.E. Cheney, Oceanographic and Geophysical Applications of Satellite Altimetry, Rev. Geoph. Space Phys. 25:875-880, 1987

Engelis, T., Global Circulation from Seasat Altimeter data, Marine Geodesy, 9, 45-69, 1985

Engelis, T., Radial Orbit error reduction and sea surface topography determination using satellite altimetry, Report 377, Ohio State University, Department of Geodetic Science and Surveying, Columbus, Ohio, 43210., 1987

Flierl G.R. and A.R.Robinson, On the time-dependent meandering of a thin jet, J.Phys.Oceanogr.14, 412-423, 1984

Joyce T.M., C. Wunsch and S.D. Pierce, Synoptic Gulf Stream Velocity Profiles through simultaneous inversion of Hydrographic and acoustic doppler data, J. Geophys. Res., 91 (C6), 7573-7585, 1989

Kelly, Kathryn A and Sarah T. Gille, Gulf Stream Surface Transport and Statistics from the Geosat Altimeter, J. Geophys. Res., in press, 1989

Marsh, J.G, F.J.Lerch, C.J.Koblinsky, S.M.Klosko, J.W. Robbins, R.G.Williamson, and G.B. Patel, Dynamic Sea Surface Topography, Gravity and Improved Orbital accuracies from the direct evaluation of Seasat Altimetry, NASA Technical Memorandum, available from Goddard Space Flight Center, Greenbelt, MD 20771,, 1989

Richardson, P.L., The Benjamin Franklin and Timothy Folger charts of the Gulf Stream, In: Oceanography, The Past. M Sears and D. Merriman (ed.), Springer Verlag, 812pp, 1980

Roemmich D. and C. Wunsch, On combining satellite altimetry with hydrographic data, J. Mar. res. 40, 605-619, 1982

Tai, C. K. and C. Wunsch, An Estimate of Global Absolute Dynamic Topography, J. Physical Oceanography, Vol 14, No. 2: 457-463, 1984

Tai, C.-K., Estimating the Basin -Scale Ocean Circulation from Satellite Altimetry. Part I: Straightforward Spherical Harmonic Expansion, J. Phys. Oceanog., 18, 1398-1413, 1988

Tapley, B.D., R.S. Nerem, C.K. Shum, J.C.Ries and D.N.Yuan, Determination of the

general ocean circulation from a joint gravity field solution, Geophys. Res. Let., 15, 1109-1112, 1988
Vazquez, Jorge, V. Zlotnicki and L.-L. Fu, Sea Level Variabilities in the Gulf Stream between Cape Hatteras and 50ºW - A Geosat study, J. Geophys. Res., submitted, 1989
Wagner C.A. , Accuracy estimates of geoid and ocean topography recovered jointly from satellite altimetry, J. geophys. Res 91, 453-461, 1986
Worthington, L.V., On the North Atlantic Circulation, Johns Hopkins University Press, Baltimore Md., 110 pp, 1976
Wunsch, Carl and E.M.Gaposhkin, On using satellite altimetry to determine the general circulation of the oceans, with application to geoid improvement, Rev.Geophys. Space Phys., 18, 725-745, 1980
Zlotnicki V., L.-L. Fu, and W. Patzert, Seasonal Variability in Global Sea Level observed with Geosat altimetry, J. Geophys. Res. in press., 1989
Zlotnicki, V. and J.G. Marsh, Altimetry, ship gravimetry and the general circulation of the North Atlantic, Geophys. Res. Lett., in press., 1989
Zlotnicki, Victor, On the Accuracy of Gravimetric Geoids and the Recovery of Oceanographic Signals from Altimetry, Marine Geodesy, 8: 129-157., 1984

STRATEGIES FOR SOLVING THE VERTICAL DATUM PROBLEM USING TERRESTRIAL AND SATELLITE GEODETIC DATA

B. Heck
Dept. of Geodetic Science, University of Stuttgart
Keplerstr. 11, D-7000 Stuttgart 1, F.R.G.

R. Rummel
Faculty of Geodesy, Technical University of Delft
Thijsseweg 11, NL-2629 JA Delft, The Netherlands

INTRODUCTION

The classical procedure of establishing precise height networks is based upon geodetic levelling, potentially including gravity information along the levelling lines. Since levelling is a relative operation some vertical datum must be fixed in order to determine "absolute" heights of benchmarks. In most cases the vertical datum of a height network has been defined by assigning zero height to the long-term mean value of local sea level observed at a fundamental tide gauge station. The vertical datum of largely extended height networks has often been fixed by employing several tide gauge stations situated along the coastline. In any case the definition of datum of classical vertical networks is connected with the concept of local mean sea level; the equipotential surface of the earth's gravity field passing through the fundamental tide gauge mark is the reference surface of heights derived from levelling.

Comparing the vertical datum definitions referring to several unconnected height networks the reference equipotential surfaces generally are not consistent. This discrepancy is caused by the fact that mean sea level is not exactly an equipotential surface of the earth's gravity field. The vertical separation between mean sea level (with respect to a specific time period) and some global equipotential surface of the gravity field is called sea surface topography (SSTop); this equipotential surface - usually called the geoid - must be defined operationally. The SSTop is generally decomposed into a permanent or quasi-permanent part being constant within the time period under consideration, and time-dependent secular or periodical variations. Some possible causes of these phenomena have been discussed in the oceanographic literature, see e.g. (Chelton and Enfield, 1986). The most important features of oceanographically derived spatial and temporal variations of SSTop are a quasi-permanent signal on the order of 1-2 m (attributed to stationary ocean circulation) and a long-term global rise of mean sea level on the order of 1-2 mm/yr (attributed to post-glacial melting and possibly the greenhouse effect). Short-periodic tidal and mesoscale variations can be filtered out by observing over a longer time period. Additional problems in defining local mean sea level are induced by vertical motions of tide gauge stations, e.g. caused by post-glacial rebound.

Due to the existence of SSTop and the procedure of vertical datum definition the heights of bench marks involved in different vertical datum zones refer to different equipotential surfaces, the separations amounting up to 1-2 m globally. In order to establish a global vertical datum with a precision on the dm- or even cm-level these discrepancies must be derived from additional observational data. In the present contribution several proposals aiming at the solution of the vertical datum problem using terrestrial and satellite geodetic data are reviewed. Special emphasis is given to an approach based upon an extended formulation of the geodetic boundary value problem (GBVP); this procedure has recently been elaborated in detail by Rummel and Teunissen (1988).

PRACTICAL SIGNIFICANCE OF A GLOBAL VERTICAL DATUM AND ITS RELATION TO THE DEFINITION OF THE GEOID

Before the seventies the geodetic community was scarcely aware of the discrepancies in vertical datums. In fact the discussion about this subject has been pushed by oceanographers comparing oceanographically derived sea level slopes along coast lines with the results of geodetic levelling. A summary of the early investigations can be found in (Fisher, 1978). In recent years the vertical datum problem has been taken up before the background of global monitoring of mean sea level and the use of space techniques like satellite positioning and altimetry, see e.g. (Rapp, 1983; Colombo, 1985).

There are several arguments indicating the practical significance of global vertical datum definition, arranged according to their degree of importance from weak to strong arguments:

1. Comparison between the results of geodetic levelling and oceanographic procedures for determining sea surface slopes over large distances requires a consistent vertical datum system along the whole coastline segment under consideration.
2. Referencing heights to different level surfaces in different vertical datum zones causes systematic regional biases in gravity anomaly data banks; the gravity anomalies refer to the level surfaces through the fundamental tide gauge station rather than to one unique geoidal surface.
3. A globally consistent system of calibration points is required for future satellite altimetry and gradiometry missions. These methods, in turn, offer tremendous prospects to dynamic oceanography.
4. Monitoring sea level changes on various time and space scales will become an important challenge for geodesy and oceanography in near future. These studies will have a large impact on the understanding of the climate/ocean interaction as well as practical consequences for coastal safeguard in river delta and low-land regions. Large-scale regional or global monitoring requires a unique, internally consistent vertical datum.

Obviously for some of the items mentioned above a relative global vertical datum will be sufficient in the sense that only height datum connections in the form of potential differences between the fundamental regional datum points are required. In contrast, the definition of an absolute vertical datum must be referred to a unique global equipotential surface, the geoid. In order to avoid confusion this term should be understood in the following sense:

The geoid is an equipotential surface of the (downward continued external) gravity field of the earth, having been reduced for the gravitational effects of luni-solar and atmospheric masses and for short-periodic variations of the earth's gravity field. Due to the existence of long-term gravity field variations the geoid must be referred to a specific time epoch.

There is no unique definition of the geoid being a (quasi-stationary) equipotential surface of the corresponding reduced gravity field; in principle any level surface in the neighbourhood of mean sea level may be chosen. Many definitions of "the geoid" can be found in the geodetic and oceanographic bibliography, the most important ones being summarized below:

1. A physical definition results from equating the constant geoidal potential W_0 with an a priori given absolute potential value U_0, e.g. the constant normal potential at the surface of a level ellipsoid with specified parameters a (length of semi-major axis), J_2 (dynamical form factor), ω (angular velocity of earth rotation), and μ = GM (geocentric gravitational constant including the mass of the atmospheric). Since the gravity potential W is (hopefully!) a unique function of position, e.g. of the geodetic-geographic latitude φ, longitude λ and ellipsoidal height h, the relation

$$W(\varphi,\lambda,h) = W_0 \qquad (1)$$

has to be solved for the ellipsoidal height h being dependent on the variables φ, λ. In order to apply the physical definition of the geoid the behaviour of the gravity potential in the spatial neighbourhood of the geoidal surface must be known beforehand in detail. Since this assumption is not fulfilled with a reasonable degree of precision the practical realization of this definition must be questioned.

2. In the framework of an oceanic definition the geoid is characterized as that level surface in relation to which the average permanent SSTop is zero when sampled globally over all oceanic regions (Mather, 1978a). This definition is based upon the application of satellite altimetry on one hand and a combination of terrestrial gravity and satellite tracking data on the other hand. Theoretically it is also possible to include tide gauge data, but only after applying corrections for disturbing effects induced by local coastal phenomena.

3. The background of another definition of the geoid is the GBVP: The geoid is understood as that level surface in relation to which the average quasi-stationary SSTop has a vanishing zero degree harmonic in solutions of the GBVP (Mather, 1978a). This approach is based on satellite altimetry data in oceanic areas and terrestrial gravimetry and levelling in land areas. The practical application of this definition is hampered by the fact that gravity anomalies in land areas are affected by the influences of vertical datum inconsistencies; another drawback are other zero-degree effects induced by the

practically applied procedures for evaluating the solution formulae of the GBVP.

4. The geoid may also be characterized by a geodetic definition: The geoid is defined as that level surface in relation to which the average quasi-stationary SSTop has zero mean as sampled at all regional levelling datums on the globe (Mather, 1978a). Another version of the geodetic definition is the following (Lelgemann, 1977): The geoid is that level surface in relation to which the (weighted) sum of squares of SSTop at all regional levelling datums takes a minimum. This definition implies that sea gravimetry refers to the geoid which is however unknown at this stage of evaluation, requiring an iterative solution procedure.

Conceptually all these definitions should be equivalent on a level of precision of a few meters. From a practical point of view the second (oceanic) definition should be preferred.

In addition to the problems mentioned above some other difficulties affect the definition of the geoid when aiming at a level of precision between 1 cm and 1 dm. Excluding all atmospherical masses has a total effect on geoidal heights on the order of 5 m. This large influence is counteracted by re-arranging the total atmospherical mass in the centre of the earth coinciding with the origin of the coordinate system. Applying the atmospherical reduction procedure proposed by IAG (1980) the indirect effects on the geoid can mostly be neglected from a practical point of view (Christodoulidis, 1976). Another problem is the permanent, non-periodical contribution of luni-solar tides consisting of a direct and an indirect effect caused by the deformation of the solid earth (for a recent discussion see e.g. Rapp, 1989). Since these effects amount up to 2-3 dm in geoidal heights the definition of the corresponding reduced gravity field has to be considered carefully.

GEODETIC DATA SOURCES FOR SOLVING THE VERTICAL DATUM PROBLEM.

In order to solve the vertical datum problem data must be available which are sensitive to the vertical datum inconsistencies. The most complete presently existing global data base is gravity anomaly data as a result of processing levellings and gravity observations and to some extent horizontal positions derived from terrestrial triangulations. Free-air gravity anomalies over land and ocean areas are influenced not only by vertical datum inconsistencies but also by deviations between the local gravity and horizontal datums with respect to the unique global datums. Additional systematic effects are produced by using the common approximate free-air reduction formula and by inconsistencies in the types of height systems; for a discussion of these systematic error sources see (Heck, 1989b). Since these systematic effects are as large or even larger than the signal caused by vertical datum inconsistencies (Laskowski, 1988) existing gravity anomaly data banks must be revised and corrected for those effects in the sense of Mather et al. (1976) before this data can be applied for solving the vertical datum problem.

Other data suitable for the present purpose are positions fixed by space methods like SLR and VLBI, geopotential models in the form of spherical harmonic coefficients derived from satellite tracking, and satellite altimetry data (GEOS-3, SEASAT, GEOSAT). Simultaneous processing of these different types of data results in estimates of sea surface heights and

SSTop (Engelis, 1987; Marsh et al., 1989; Schrama, 1989). Geographically correlated systematic errors up to a few decimeters - caused by radial orbit errors of altimeter satellites - may still be inherent to this data source.

The future situation will be characterized by more data of type threedimensional position fixes, by improved spherical harmonic coefficients sets through dedicated gravity field missions (GRM, ARISTOTELES), and new highresolution satellite altimetry data (e.g. ERS-1, TOPEX/POSEIDON). As a final goal all these data sources should be combined in order to approach a practical solution of the vertical datum problem.

A SURVEY OF STRATEGIES FOR SOLVING THE VERTICAL DATUM PROBLEM

Since the seventies a number of approaches have been proposed in geodetic literature tackling the vertical datum problem. In the following the basic ideas of the most important (from a subjective personal view) procedures are summarized, discussing in more detail an approach recently described by Rummel and Teunissen (1988).

Oceanographic Approaches

The starting point of the oceanographic approaches is Newton's second theorem

$$\ddot{\underset{\sim}{x}} = \underset{\sim}{\kappa} \tag{2}$$

the specific force $\underset{\sim}{\kappa}$ being composed in the following way (Mather, 1976; Bosch, 1985; Rummel, 1985):

$$\underset{\sim}{\kappa} = \text{grad } V - \frac{1}{\rho} \text{grad } p + \underset{\sim}{F} \tag{3}$$

where V is the instantaneous gravitational potential, ρ mass density, p hydrostatic pressure, and $\underset{\sim}{F}$ represents all disturbing forces such as tidal forces and wind friction. Equation (2) holds for the position vector $\underset{\sim}{x}$ of a moving particle being described with respect to an inertial reference frame. A more convenient description is given with respect to a moving frame rotating with the earth with (constant) angular velocity ω. Introducing a topocentric local level system with orthonormal base vectors $\underset{\sim}{g}_i$ ($\underset{\sim}{g}_3$ vertical, $\underset{\sim}{g}_2$ north, $\underset{\sim}{g}_1$ east), $i \in \{1,2,3\}$

$$\underset{\sim}{x} = x_i \underset{\sim}{g}_i \quad , \quad \underset{\sim}{F} = F_i \underset{\sim}{g}_i \tag{4}$$

and defining the local gravity vector $\underset{\approx}{g}$

$$\underset{\approx}{g} = -g \, \underset{\sim}{g}_3 \tag{5}$$

as the sum of centrifugal and accelerational components the following set of differential equations is obtained:

$$\ddot{x}_1 + 2\omega(\dot{x}_3 \cos\varphi - \dot{x}_2 \sin\varphi) = -\frac{1}{\rho}\frac{\partial p}{\partial x_1} + F_1 \tag{6a}$$

$$\ddot{x}_2 + 2\omega\dot{x}_1 \sin\varphi = -\frac{1}{\rho}\frac{\partial p}{\partial x_2} + F_2 \tag{6b}$$

$$\ddot{x}_3 - 2\omega\dot{x}_1 \cos\varphi = -g - \frac{1}{\rho}\frac{\partial p}{\partial x_3} + F_3. \tag{6c}$$

These relationships can be simplified using some reasonable approximations (Mather, 1976) and inserting the Coriolis parameter f

$$f := 2\omega \sin\varphi \tag{7}$$

where φ is the ellipsoidal latitude

$$\ddot{x}_1 - f\dot{x}_2 = -\frac{1}{\rho}\frac{\partial p}{\partial x_1} + F_1 \tag{8a}$$

$$\ddot{x}_2 + f\dot{x}_1 = -\frac{1}{\rho}\frac{\partial p}{\partial x_2} + F_2 \tag{8b}$$

$$0 = -g - \frac{1}{\rho}\frac{\partial p}{\partial x_3} + F_3. \tag{8c}$$

These formulae are the background of the oceanographic procedures of steric levelling and dynamic (or geostrophic) levelling. Setting in equation (8c) $F_3 \approx 0$ and $g = -\partial W/\partial x_3$ and integrating the resulting expression the basic equation of steric levelling is obtained

$$W_2 - W_1 = \int_{p_1}^{p_2} \frac{1}{\rho}\, dp \quad . \tag{9}$$

By equation (9) the potential difference between the ocean surface (level 2) and a reference level (level 1) can be evaluated from pressure, temperature and salinity profiles. Implicitly it is assumed that the reference level $W = W_1$ is an isobaric surface or "surface of no motion". This assumption can be expected to be valid with a reasonable degree of precision at most in deep ocean areas. In addition, equation (9) presupposes also the sea surface to be isobaric; this assumption can be circumvented by applying corrections for horizontal pressure gradients. The solution of equation (8c) can also be represented in terms of estimates of currents.

Equations (8a) and (8b) form the basis of dynamic (geostrophic) levelling. These formulae are suitable for extrapolating current and SSTop information across shallow coastal waters. The precision of extrapolated SSTop is mainly restricted by modelling friction forces.

Due to the inadequacy of hypotheses and imperfect modelling of frictional forces the oceanographic procedures of steric and geostrophic levelling are not directly applicable to establish vertical datum connections between continents with the intended precision.

Combination of satellite altimetry and dynamic levelling

An approach based upon satellite altimetry and geostrophic levelling has been proposed by Mather (1976). According to this proposal the SSTop between two continental areas is determined from satellite altimetry data and geopotential models as derived from satellite tracking. These SSTop data are then extrapolated over the shallow continental shelf areas to the fundamental tide gauges using dynamic levelling (formulae 8a,b). As a result the differences between SSTop at the tide gauge stations situated on various continents are obtained. This procedure is affected by the same difficulties as mentioned above, i.e. by the problem of adequate modelling of frictional forces in continental shelf areas (e.g. bottom friction and surface wind stress).

Datum connection by gravimetry and satellite positioning

The approach elaborated by Colombo (1980a,b, 1985), Rapp (1980) and Hajela (1985) aims at the determination of potential differences among a set of bench marks situated on various continents. The following data configuration is prescribed:

Three-dimensional geocentric position fixes from Satellite Laser Ranging at a number of bench marks P_i and Q_j, bench marks P_i being situated in the datum zone "A" and Q_j in the datum zone "B".

Potential differences $\Delta W(BMA,P_i)$ between bench marks P_i and a fundamental datum point BMA in the datum zone A, and potential differences $\Delta W(BMB,Q_j)$ between bench marks Q_j and a fundamental datum point BMB in the datum zone B, respectively.

Some high-degree spherical harmonic expansion of the gravity potential representing the global, long and medium wavelength part of the geopotential (e.g. up to degree 180 or 360).

Local gravity surveys in a cap around every bench mark P_i and Q_j, respectively, controlling the local part of the potential not covered by the geopotential model.

From this data so-called "modified" gravity anomalies can be derived which are biased by the unknown vertical datum inconsistency. In the final step the potential difference $\Delta W(BMA,BMB)$ among the fundamental datum points BMA and BMB is estimated using the least squares collocation algorithm. Formal accuracy estimates from collocation (Hajela, 1985) point out that the vertical datum connection between USA and Western Europe is possible with a precision of about 0.5 kgalm. Further improvements will be feasible with the advent of improved geopotential models and inclusion of more position fixes, e.g. by relative GPS positioning.

Geodetic boundary value problem approach.

Rummel and Teunissen (1988) discussed an approach based upon an extended formulation of the geodetic boundary value problem (GBVP). The data included in this procedure are potential differences in the form of geopotential numbers (or orthometric or normal heights) and gravity anomalies as they are applied in practice. Strictly speaking the formulation according to the scalar free gravimetric GBVP is used assuming the horizontal coordinates of the boundary points to be known e.g. from satellite positioning or terrestrial triangulation (Sacerdote and Sansò, 1986; Heck, 1989a). Linearizing the boundary conditions for scalar gravity g and geopotential numbers C with respect to some fixed normal potential and approximate boundary surface (telluroid) the boundary equations in linear and spherical approximation take the following form:

$$\Delta g = -\frac{2\gamma}{r} \cdot \zeta - \frac{\partial T}{\partial r} \tag{10}$$

$$\Delta c = \gamma \cdot \zeta - T + \Delta\tilde{w}. \tag{11}$$

The gravity anomaly Δg is the difference between actual gravity at an arbitrary surface point and normal gravity at the corresponding telluroid point; the geopotential number anomaly Δc is defined equivalently. The unknowns in equations (10), (11) are the height anomaly ζ (vertical separation between associated surface and telluroid points) and the disturbing potential T which is assumed to be harmonic in the space outside the masses and regular at infinity. In addition a term $\Delta\tilde{w}$ has been included in (11) due to the unability to observe absolute potential values. This term essentially describes vertical datum effects. According to the assumptions put into this term several versions of the extended GBVP can be set up. In the following the solution formulae are presented without derivation; for details see (Rummel and Teunissen, 1988).

Model 1: All Vertical Datum Zones Connected with One Global Fundamental Site. In this case $\Delta\tilde{w} = \Delta w_0$ is constant and describes the unknown bias in the potential W under the restriction that the gravitational part in W vanishes at infinity; as a consequence T is forced to be regular at infinity. The solution of this problem is given in the following analytical form:

$$T = 2\frac{R}{r} \cdot \Delta w_0 + \left(\frac{R}{r}\right)^2 \cdot T_1 + \\ + \frac{R}{4\pi} \iint_\sigma (\Delta g + \frac{2}{R}\Delta c) \cdot [S(r,\psi) - \frac{R}{r}] \cdot d\sigma + \ldots \tag{12}$$

$$\zeta = \frac{1}{\gamma}(\Delta c + T - \Delta w_0) \\ = \frac{\Delta c}{\gamma} + \frac{\Delta w_0}{\gamma} + \frac{1}{\gamma} \cdot T_1 + \\ + \frac{R}{4\pi\gamma} \iint_\sigma (\Delta g + \frac{2}{R}\Delta c) \cdot [S(\psi) - 1] \cdot d\sigma + \ldots \tag{13}$$

Higher order terms expressing the influence of topography have been omitted. R is a mean radius of the earth, γ denotes the mean value of normal gravity over the surface. $S(r,\psi)$ and $S(\psi)$ are Stokes-Pizzetti and Stokes functions, respectively. Obviously for evaluating (12) and (13) the anomalies Δg and Δc must be given all over the boundary surface; if a Marussi type mapping $\Delta c \equiv 0$ is applied for defining the telluroid only gravity anomalies Δg will remain.

The right hand sides of (12), (13) contain four degrees of freedom, corresponding to three free parameters in the first-degree surface spherical harmonic term T, and the constant Δw_0. This indeterminacy of the solution cannot be removed unless additional observational data is included, e.g. the absolute geocentric position of one site and the vertical position (in geometrical sense) of another site. Setting conventionally $T_1 \equiv 0$, $\Delta w_0 \equiv 0$ will give results which are generally not consistent with observational data of other type such as e.g. satellite positioning and terrestrial distance measurements.

Model 2: K + 1 Different Vertical Datums Covering the Whole Globe. Weakening the presuppositions it will now be assumed that there are K + 1 different datum zones covering the whole globe. In any date zone $\Delta\tilde{w}$ takes a constant value, e.g. in the k-th zone $\Delta\tilde{w} = \Delta w_k$, $k \in \{0,1,\ldots,K\}$. Referring an arbitrary datum zone bias Δw_k to one basic reference datum, e.g. k = 0, the datum bias Δw_i, $i \in \{1,\ldots,K\}$ is splitted in two parts

$$\Delta w_i = \Delta w_0 + C_{0Q_i} \tag{14}$$

$$C_{0Q_i} := W(Q_i) - W(0) \tag{15}$$

where 0 can be considered as the basic fundamental bench mark and Q_i the fundamental bench mark of i-th datum zone. The solution formula for ζ is now (assuming for simplicity (r = R)):

$$\zeta = \frac{\Delta c}{\gamma} + \frac{\Delta w_0}{\gamma} + \frac{1}{\gamma} \cdot T_1 - \frac{1}{\gamma} C_{0Q_i} +$$

$$+ \frac{R}{4\pi\gamma} \iint_\sigma \left[\Delta g + \frac{2}{R}\Delta C - \frac{2}{R} C_{0R_j}\right] \cdot [S(\psi) - 1] \cdot d\sigma + \ldots \tag{16}$$

containing K + 4 degrees of freedom (K constants C_{0Q_i} referring to the datum zones i = 1 to i = K, respectively, Δw_0 and three constants in T_1). These degrees of freedom in the solution can be balanced by the absolute geocentric position of one site and the vertical positions of K + 1 other sites.

Obviously the effect of the vertical datum biases on ζ is two-fold: There is a direct effect through the terms $(\Delta w_0 - C_{0Q_i})/\gamma$, and there is an indirect effect through the integral term, the integration extending over any datum zone $j \in \{1,2,\ldots,K\}$. Since C_{0Q_j} is constant over an arbitrary datum zone this integral can be written in the following way:

$$- \frac{1}{2\pi\gamma} \iint_\sigma C_{0Q_j} \cdot [S(\psi) - 1]d\sigma =$$

$$= - \frac{1}{2\pi\gamma} \sum_{j=1}^{K} C_{0Q_j} \iint_{\Delta\sigma_j} [S(\psi) - 1]d\sigma \quad , \tag{17}$$

$\Delta\sigma_j$ denotes the area of the unit sphere covered by the j-th datum zone.

Model 3: K + 1 Different Datum Zones over Land, SSTop in Oceanic Areas. The solution of this most general, realistic case can be expressed in the following way:

$$\zeta = \frac{\Delta c}{\gamma} - \frac{\Delta\tilde{w}}{\gamma} + \frac{1}{\gamma} \cdot T_1 +$$

$$+ \frac{R}{4\pi\gamma} \iint_\sigma [\Delta g + \frac{2}{R}\Delta c - \frac{2}{R}\Delta\tilde{w}][S(\psi) - 1] \cdot d\sigma + \ldots \tag{18}$$

The function $\Delta\tilde{w}$ comprises the influence of SSTop ζ_s over ocean areas, the relationship between $\Delta\tilde{w}$ and ζ_s being given by

$$\Delta\tilde{w} = \Delta w_s - \gamma\zeta_s \qquad \text{(in ocean areas).} \tag{19}$$

The constant Δw_s denotes the unknown difference between the absolute gravity potential value referred to the equipotential reference surface for SSTop heights ζ_s ("oceanic geoid") and the nominal potential value accepted for this surface.

Over land areas $\Delta\tilde{w}$ has the same form as described in equation (14)

$$\Delta\tilde{w} = \Delta w_k = \Delta w_0 + C_{0Q_k} \quad , \qquad k \in \{0,1,\ldots,K\} \tag{20}$$

(in land areas, k-th datum zone).

Under these promises $\Delta\tilde{w}$ is a piecewise continuous function on σ. If the SSTop is represented in analytical form (e.g. by a finite degree surface spherical harmonic expansion) the function $\Delta\tilde{w}$ is analytical over the ocean areas.

Obviously the SSTop acts upon the height anomaly ζ in a direct and an indirect way via the integral in (18). In (Rummel and Teunissen, 1988) it is shown that both direct and indirect effects have the same order of magnitude; the total effect ranges from -1.8 m to +1.2 m.

Remark: The form (20) of $\Delta\tilde{w}$ over land areas can even be generalized to include a description of systematic levelling errors in any datum zone. The form of the solution formula (18) will remain the same. In this extended form $\Delta\tilde{w}$ over the k-th datum zone can be represented in functional form (e.g. surface spherical harmonic expansion) involving a set of unknown parameters.

Based upon model 3 the vertical datum bias parameters C_{0Q_k}, Δw_0 and eventually Δw_s can be derived simultaneously with the solution of the GBVP. Generally Δc is assumed to be zero, corresponding to a Marussi type telluroid mapping. The following data configuration is prescribed:

Gravity anomalies over land areas referred to a specific vertical datum in any datum zone $k\in\{0,1,\ldots,K\}$. The gravity anomaly data should be corrected for all systematic terms other than those induced by vertical datum inconsistencies.

Gravity anomalies over ocean areas taking the SSTop into account; ζ_s must be given with respect to some equipotential surface (oceanic geoid).

Three-dimensional geocentric position fixes for at least K + 2 sites, at least one site situated in each datum zone, derived by space methods.

Normal (or orthometric) heights at the sites where three-dimensional position fixes are available, derived from levellings and gravity observations along the levelling lines. These normal heights are related to the Marussi type telluroid mapping and refer to the respective datum zones.

Non-terrestrial data in form of three-dimensional position fixes are introduced in order to solve for the unknown vertical bias parameters and the first-degree potential term T_1. Combining three-dimensional position fixes with normal heights additional information about the height anomalies ζ is obtained which can be used for determining T_1 and $\Delta\tilde{w}$. If more than K+2 position fixes are available a least-squares approach can be applied, (Rummel and Teunissen, 1988).

As a result the procedure proposed by Rummel and Teunissen (1988) enables a simultaneous determination of the disturbing potential, height anomalies, and datum parameters describing vertical datum biases and the first-degree spherical harmonic component T_1. The approach proposed by Mather (1978b), Mather and Rizos (1978) and Mather et al. (1978) shows some similarities to the procedure described above and could be regarded as its forerunner.

CONCLUSIONS

Vertical datum connections between different continents on the basis of mean sea level approximating a global equipotential surface have a precision not better than 1-2 m due to the existence of SSTop.

An improved precision in the order of 10 cm seems to be attainable combining several types of data:

Gravity observations and levelling data which can be transformed into gravity anomalies referred to the vertical datum of the respective datum zone. There is a strong need for globally consistent gravity anomaly data banks screened for all systematic effects other than those induced by vertical datum inconsistencies. Sea surface topography information in order to reduce marine gravimetry and to remove systematic long wavelength errors from oceanic gravity anomalies.

Three-dimensional geocentric position fixes precise to about 5 cm, at least at one site in each datum zone; one more absolute position fix is required in order to solve for the first-degree terms in the spherical harmonic expansion of the disturbing potential.

Global geopotential models of adequate precision; further improvements in this direction can be expected in the near future through dedicated satellite missions for gravity field determinations, e.g. using satellite gradiometry.

Based upon these data it can be expected that vertical datum connections among different continents can be performed in future with a precision equivalent to the presently achieved accuracy of horizontal datum transformations.

Acknowledgements. The first author has been supported by the German Research Foundation (Deutsche Forschungsgemeinschaft) in the framework of a Heisenberg Research Grant.

REFERENCES

Bosch, W. (1985) Concept for modelling the sea surface topography, *I. Hotine-Marussi Symposium on Mathematical Geodesy,* 787-807, Rome.

Chelton, D.B., D.B. Enfield (1986) Ocean Signals in Tide Gauge Records, *Journ. Geoph. Res.,* **91**, B9, 9081-9099.

Christodoulidis, D.Z. (1976) On the realization of a 10 relative oceanic geoid, *Dept. of Geod. Sci. Rep.* **247**, Ohio State Univ., Columbus.

Colombo, O.L. (1980a) A world vertical network, *Dept. of Geod. Sci. Rep.* **196**, Ohio State Univ., Columbus.

Colombo, O.L. (1980b) Transoceanic Vertical Datum Connections, *Proc. Symp. on Problems Related to the Redefinition of the North American Vertical Geodetic Networks*, 87-101, Ottawa.

Colombo, O.L. (1985) Levelling with the help of space techniques, *Proc. 3rd Int. Symp. on the North Americal Vertical Datum*, NOAA, Rockville, MD.

Engelis, T. (1987) Radial orbit error reduction and sea surface topography determination using satellite altimetry, *Dept. of Geod. Sci. Rep.* **377**, Ohio State Univ., Columbus.

Fischer, I. (1978) Mean sea level and the marine geoid - an analysis of concepts, *Marine Geodesy*, **1**, 37-59.59.

Hajela, D.P. (1983) Accuracy estimates of gravity potential differences between Western Europe and United States through LAGEOS Satellite Laser Ranging Network, *Dept. of Geod. Sci. Rep.* **345**, Ohio State Univ., Columbus.

Heck, B. (1989a) A contribution to the scalar free boundary value problem of physical geodesy, *manuscripta geodaetica*, **14**, 87-99.99.

Heck, B. (1989b) An evaluation of some systematic error sources affecting terrestrial gravity anomalies, *Bull. Géod.*, in press.

IAG (1980) The Geodesist's Handbook, *Bull. Géod., 54*, No. **3**.

Laskowski, P. (1983) The effect of vertical datum inconsistencies on the determination of gravity related quantities, *Dept. of Geod. Sci. Rep.* **349**, Ohio State Univ., Columbus.

Lelgemann, D. (1977) On the definition of the Listing geoid taking into consideration different height systems, *Nachrichten Karten-u. Vermessungswesen, Rh. II*, **34**, 25-47.

Marsh, J.G. et al. (1989) Dynamic sea surface topography, gravity, and improved orbit accuracies from the direct evaluation of SEASAT altimeter data, *NASA Techn. Memo*, **100735**, GSFC, Greenbelt, Md.Md.

Mather, R.S. (1976) Some possibilities for recovering oceanographic information from the SEASAT missions. *Unisurv*, **G24**, 103-122.22.

Mather, R.S. (1978a) The role of the geoid in four-dimensional geodesy, *Marine Geodesy*, **1**, 217-252.52.

Mather, R.S. (1978b) The geoid and continental gravity data banks: the role of satellite altimetry, *Unisurv*, **G29**, 1-9.-9.

Mather, R.S., Rizos, C., Hirsch, B. and Barlow, B.C. (1976) An Australian gravity data bank for sea surface topography determinations (AUSGAD 76), *Unisurv*, **G25**, 54-84.

Mather, R.C. and Rizos, C. (1978) On vertical datum definition from GEOS-3 altimetry, *Proc. 2nd Int. Symp. on Problems Related to the Redefinition of North American Geodetic Networks*, Arlington, 589-597.

Mather, R.S., Rizos, C. and Morrison, T. (1978) On the unification of geodetic levelling datums using satellite altimetry, *NASA Techn. Memo* **79533**, GSFC, Greenbelt, Md.

Rapp, R.H. (1980) The Need and Prospects for a World Vertical Datum, *Proc. Symp. on Problems Related to the Redefinition of the North American Vertical Geodetic Networks*, Ottawa.

Rapp, R.H. (1988) The need and prospects for a world vertical datum, *Proc. IAG Symposia, IUGG General Assembly*, Hamburg, vol. **2**, 432-445.

Rapp, R.H. (1989) The treatment of permanent tidal effects in the analysis of satellite altimeter data for sea surface topography, *Manuscripta Geodaetica*, in press.

Rummel, R. (1985) Satellite altimetry as part of a geodetic model, *I Hotine-Marussi Symposium on Mathematical Geodesy*, 757-787, Rome.

Rummel, R. and Teunissen, P. (1988) Height datum definition, height datum connection and the role of the geodetic boundary value problem, *Bull. Géod.*, **62**, 477-498.

Sacerdote, F. and Sansò, F. (1986) The scalar boundary value problem of physical geodesy, *manuscripta geodaetica*, **11**, 15-28.

Schrama, E.J.O. (1989) The role of orbit errors in processing of satellite altimeter data, Netherlands Geodetic Commission, New Series, 33, Delft.

VERTICAL DATUM DEFINITION FOR OCEANOGRAPHIC STUDIES USING THE GLOBAL POSITIONING SYSTEM

Chris Rizos
School of Surveying, University of N.S.W.,
Sydney, N.S.W. 2033, AUSTRALIA

Richard Coleman
School of Civil and Mining Engineering, University of Sydney,
Sydney, N.S.W. 2006, AUSTRALIA

INTRODUCTION

Australia's climate is particularly influenced by global ocean circulation as its coastlines border the Pacific and Indian Oceans and form a prominent northern boundary of part of the Southern Ocean. The magnitude of meridional heat flux and transport in these oceanic areas is poorly understood, and the seasonal and interannual variability is not known at all. Variations in these quantities can produce wide reaching climatological effects for Australia, and there is therefore a crucial need to improve our understanding of the ocean circulation in the neighbouring oceans.

One of the main forcing functions driving ocean surface circulation is the pressure gradient, or sea surface slope. This is, for all intents and purposes, equivalent to the slope of the sea surface relative to the **geoid**, an equipotential surface of the earth's gravity field about which the global Mean Sea Level (MSL) is zero (Rizos, 1980). There are a number of methods by which the sea surface (vertical) geometry can be defined:

1. A class of techniques collectively known as oceanographic levelling (Ibid, 1980). **Steric levelling** is based on the determination of variations in the density of vertical water columns, from temperature and salinity measurements. **Geostrophic levelling** relies on inverting the Equations of Motion (of fluids on a spinning body) by using observed current vectors as input. Both techniques provide the sea surface slope relative to the geoid (or more correctly the zero pressure surface).
2. Geodetic levelling (Heiskanen and Moritz, 1967) gives the **orthometric height** of points on land relative to the equipotential surface of the earth's gravity field at the vertical datum station (this may be coincident with the geoid, but is more likely to be parallel to it and offset from it by an amount of the order of a metre).

Geodetic and oceanographic levelling results can be compared at MSL tide gauges. However both oceanographic and geodetic levelling have their problems. In particular:

a. Geodetic levelling cannot span expanses of water wider than a few kilometres.

b. Steric levelling is only appropriate for deep water (greater than about 1000m).

c. Geostrophic levelling is of limited accuracy due to problems of measuring current vectors accurately.

d. The sea is in continual motion, with sea surface slopes changing both spatially and temporally, hence oceanographic levelling results are at best averages.

There are two alternative techniques based on satellite technologies:

1. The Global Positioning System (GPS) can be used to determine the relative height of two or more ground antennae, with respect to a geocentrically located ellipsoid.
2. A satellite-borne radar altimeter is capable of measuring the ocean surface height over spatial scales of 2-10km to within about 5cm accuracy. Again, the height is with respect to a geocentrically located ellipsoid.

The vertical geometry involved is illustrated in Fig. 1.

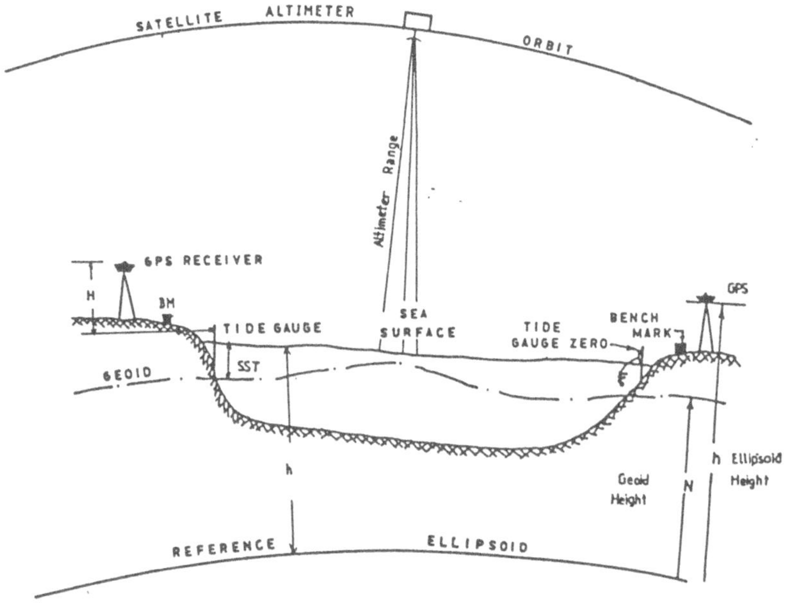

Fig. 1. Vertical geometry of tide gauge MSL, geoid, GPS heights and satellite altimetry.

What is required to link geodetic levelling and oceanographic levelling with the remotely sensed sea surface height data and GPS results is a unique vertical datum that integrates geometric information (ellipsoid heights) and gravity field contributions (geoid heights).

This report briefly describes a project presently underway to develop and test procedures for vertical datum definition adequate for oceanographic studies over regional areas. The area chosen for investigation is the Bass Strait region between mainland Australia and the island of Tasmania. Here a number of intensive oceanographic surveys have been carried out (for example, Baines and Hubbert, 1988), and there is corroborative information from 3-dimensional numerical models (Fandry et al., 1985). Fig. 2 indicates the location of some of the oceanographic sensors that have been deployed in Bass Strait (Baines and Hubbert, 1988).

Bass Strait Vertical Datum Definition Project

The main phases of the project are:

1. To define the geometry of a series of tide gauges in the Bass Strait area in relation to the uniquely defined satellite datum provided by GPS.

2. To determine the relative geoid heights at the tide gauges using the gravimetric technique described by Kearsley (1988), using available surface gravity data and global models of the gravitational field.

3. To compare sea surface slopes derived geodetically (GPS results reduced to the geoid) with those derived from an analysis of current observations (geostrophic levelling). The combination of oceanographically-derived sea surface slopes (essentially the fine structure information, but relatively weakly defined in an absolute framework) with the geodetic results that span the Bass Strait, will provide the basis for defining a unique vertical datum with a precision better than a decimetre.

4. This datum definition provides the groundwork for the calibration of the next generation of satellite radar altimeter missions - ERS-1, to be launched in 1991, and TOPEX/POSEIDON, launch planned for 1992.

This report details only the first phase - the Bass Strait GPS Survey.

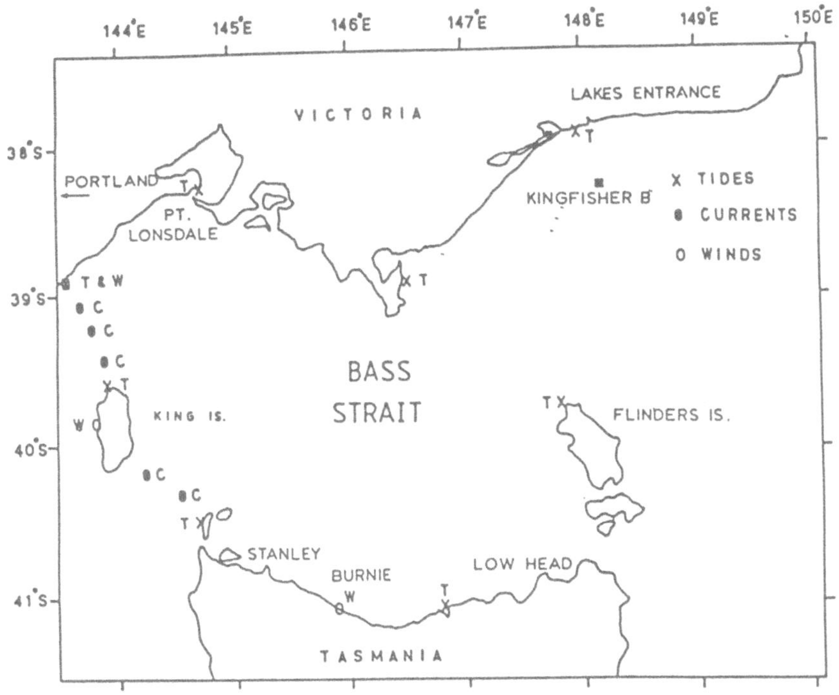

Fig. 2. The Bass Strait region showing locations of oceanographic sensors and intended GPS sites.

THE BASS STRAIT GPS SURVEY - PLANS & RATIONALE

GPS was designed primarily as a navigation system, but through the use of special measurement and analysis techniques, GPS is already having an enormous impact in the fields of surveying and geodesy (King et al., 1987). Although the GPS system is still technically in its testing phase, it has already demonstrated its capability for high precision surveying, for example in support of crustal motion surveys. GPS receivers are operated in the differential mode, in which two or more receivers track the visible GPS satellites for periods ranging from half an hour to several hours (depending on the accuracy being sought). Relative positioning accuracies of the order of 0.1 parts per million (ppm), or better, have been demonstrated. Hence, for antenna separations of the order of 100km, relative positioning accuracies of the order of a centimetre or better are possible. However, the accuracies of height determination are typically an order of 2 or 3 times worse than for the horizontal components.

This project will therefore provide the impetus to develop observation and data reduction strategies most appropriate for vertical, as opposed to horizontal, GPS surveys. A by-product of this first phase of the project will be the definition of a unified height datum for Australia. At present the Australian mainland vertical datum is unrelated to the Tasmanian height datum. This is of relevance to the geodetic community. Furthermore, this vertical datum unification by geodetic means is also of significance for monitoring possible longterm sea level fluctuations arising from the "Greenhouse Effect".

The Fieldwork

9 GPS observation sites have been identified. In the case of 7 of the sites, they are adjacent to operating tide gauges, while the remaining 2 sites are benchmarks in the vicinity of which tide gauge records had been collected in the past and for which there is a direct vertical connection from the benchmark to local MSL (obtained from the analysis of the tide gauge records). The 3 Victorian sites are at Portland, Pt. Lonsdale and Lakes Entrance. The 3 Tasmanian sites are at Stanley, Burnie and Low Head. The remaining 3 sites are located on King Island, Flinders Island and the Kingfisher B oil platform. The sites are shown in Fig. 2.

Over a 7 day period in early September, 1989, 3 dual-frequency GPS receivers will be deployed in such a manner as to provide a network of interlocking triangles connecting, directly and indirectly, all the tide gauges surrounding the Bass Strait basin. The daily observation session at each site will be 6-7 hours in length, and a minimum of 2 and a maximum of 6 GPS satellites will be tracked while they are above 15° elevation. For added strength (and in order to have a fair degree of redundancy), each site will be occupied for at least 2 sessions (days), and some for 3 sessions. In addition to the 3 GPS receivers in Bass Strait, 2 receivers will be operated continuously at sites in Canberra, on the Australian mainland, and in Hobart, in the SE of Tasmania.

Data Reduction Strategies

To achieve the relative accuracy sought in this experiment (0.1ppm), a number of important observation and data reduction requirements have to be met:

1. Some form of GPS satellite orbit adjustment has to be attempted.
2. Dual-frequency observations are essential to eliminate systematic errors arising from ionospheric refraction.
3. Appropriate tropospheric modelling will be need to be carried out.
4. Long tracking sessions, observing as many of the visible satellites as possible, are critical.

While each of these are important for any high accuracy GPS survey, the first requirement is the most critical and deserves special consideration in this project.

Orbit Adjustment Strategies. The **Fiducial Concept** is fundamental to most orbit computation experiments (e.g. Kellogg et al., 1989). In such a strategy, a network of tracking stations whose locations are accurately known from VLBI define a self-consistent coordinate frame to which all GPS orbits (and baseline) solutions are referred. An alternative to the fiducial approach – the **Free Network** approach – in which all except one site's coordinates, as well as the GPS orbits, are simultaneously adjusted is described in Beutler et al. (1986), and is perhaps more appropriate to this experiment. The Canberra site, where a continuously operating GPS receiver will be located, is at the Orroral Satellite Laser Ranging station. This site's coordinates are assumed known with respect to the geocentre at the 0.5m level. This is rather conservative, but it is the value adopted for the origin uncertainty of this station in the Australia GPS Orbit Determination Pilot Project (Rizos et al., 1989). The coordinates of the Hobart site, as well as the tide gauge sites will be free to adjust. This is the Free Network option.

Since planning for this project commenced, we have learnt that during the same time frame a large GPS campaign will be observed in S.E. Asia and S.W. Pacific (the so-called "APEX" experiment). Initial plans are to deploy continuously operating GPS receivers at 3 Australian mainland sites (Yaragadee, Darwin and Orroral) and at Wellington, New Zealand. The coordinates of these sites will be established within a Fiducial Network presently restricted to North America and Europe. We have therefore a further data reduction option of incorporating data from the 3 additional tracking stations in the region, and holding their coordinates fixed, together with those of Orroral (and hence defining the reference frame). This is the Fiducial Network option.

Another important issue relates to the length of the adjusted satellite arc. With 2 GPS receivers tracking continuously for 7 days, one 7 day arc will be adjusted, and with it different sets of site coordinates (for the sites tracking during that day). However the following analyses focus on the accuracy of the derived tide gauge site coordinates (primarily the vertical component) from one and two days of tracking, assuming the satellite orbit is adjusted for arcs of this length.

To study the impact of these data reduction strategies on the accuracy of the tide gauge site coordinates, it was decided to perform a covariance analysis study.

THE BASS STRAIT GPS SURVEY - A COVARIANCE ANALYSIS

The study of systematic errors in simulated least squares adjustments is known variously as **covariance analysis**, **sensitivity analysis** or **consider analysis** (Grant, 1989). These names arise because the **cross covariance** matrix between systematic biases and the estimated parameters is used to determine the **sensitivity** of the solution to unadjusted parameters that are **considered** to be biased.

In the case of GPS data adjustment, systematic biases have origins in the satellite orbits (force models), atmospheric refraction (troposphere and ionosphere) and the reference system (fixed station coordinates, reference frame orientation parameters, etc.). In this report we summarise the results of a covariance analysis in which the effects of reference system uncertainties and length of satellite arc on the initial (adjusted) elements of the GPS satellites are studied. The DASH software, developed at the School of Surveying, University of N.S.W., was used to simulate a simultaneous satellite orbit and tracking station adjustment using GPS phase data (Grant, 1989).

A Priori Uncertainties of Observations and Adjusted Parameters

Tracking Station Coordinates. Other than Orroral (and the Fiducial Network stations), all other sites were assumed to have coordinates derived from a standard GPS survey adjustment with fixed orbits (based on the Broadcast Ephemeris), and their relative coordinates are therefore only known to the one metre level.

Satellite Parameters. The initial elements (position and velocity) of the GPS satellites were adjusted, for 1 and 2 day arc lengths. The initial elements refer to the starting epoch of the arc.

GPS Biases. As DASH simulates a GPS phase data adjustment, cycle ambiguities and explicit satellite and receiver clock errors are included in the solve-for parameter set. We investigated the improvement (if any) in the coordinate accuracies when the ambiguities were correctly resolved.

Atmospheric Refraction. Ionospheric delay was assumed to be known without error from dual-frequency observations. Residual tropospheric delay error after the observations have been corrected *a priori* using surface met data, a tropospheric refraction model, or both, was assumed to have a constant zenith magnitude of 0.048m (approximately 2% of the full effect).

Measurement Noise. The receiver measurement error (all receiver and antenna errors) was assumed to be 0.004m for 4 minute data.

The *a priori* uncertainties are listed in Table 1.

The Systematic Error Models

In addition to defining observation scenario and specifying the *a priori* uncertainties of the observations and adjusted parameters, *a priori* uncertainties of the unadjusted parameters need also to be specified.

Tracking Station Coordinates. The coordinates of Orroral were held fixed, and considered free from error, hence they define the reference frame origin. In the Fiducial Network option the uncertainties in the coordinates of Yaragadee, Darwin and Wellington were considered as sources of systematic errors.

Reference Frame. The uncertainty in the coordinates of the geocentre (about which the satellites orbit) in the reference frame defined by the fixed Orroral tracking station coordinates was assumed to be an additional source of systematic error. Uncertainties in polar motion, UT1 and GM were also included.

Satellite Perturbation Model. Uncertainties in the gravity field model and from residual solar radiation pressure (it was assumed to be modelled with an uncertainty of 5% of its full effect) were also considered.

The uncertainties in these unadjusted parameters are listed in Table 1.

Table 1. *A priori* uncertainties of observations, adjusted and unadjusted parameters in the simulation of a combined orbit / network adjustment for Bass Strait baselines.

<u>OBSERVATIONS</u>	
4 minute data	0.004m
<u>ADJUSTED PARAMETERS</u>	
Receiver Coordinates	
Survey stations (Hobart and Bass Strait sites)	1.0m (east, north, height)
Orbits: Initial elements	20m (5m, 12m, 15m HCL)
<u>UNADJUSTED PARAMETERS</u>	
Receiver Coordinates	
Origin (Orroral)	0m by definition
Tracking stations (Fiducial net option)	0.02m (east, north, height)
Reference Frame	
Centre of mass	0.5m (x, y, z)
Polar motion, UT1	0.01 arcsec (0.3m)
GM	0.02 ppm
Force model	
Gravity field	100% of GEM-9 to (8,8)
Solar radiation pressure	5% of effect
Troposphere	
Constant zenith delay	0.048m (2% of effect)

The Results

The accuracies of the coordinate components of a typical 130km baseline between tide gauges in Bass Strait assuming the Free Network option are shown in Fig. 3, for various arc lengths and for the case where ambiguities are resolved (ambiguity-fixed solution) or left free. The total error is the RSS ("root sum squared") of the individual errors for each position component.

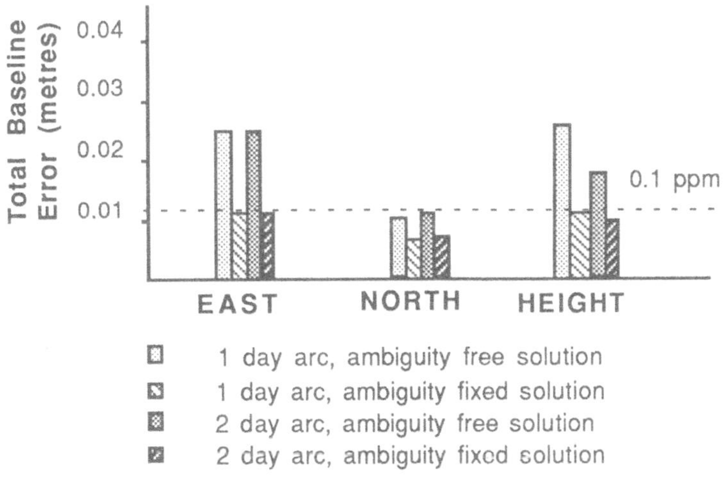

Fig. 3. Accuracy of 130km baseline for 6hr data session(s), assuming Free Network option (Orroral site fixed, Bass Strait sites free).

A breakdown of the error sources is given in Fig. 4, for the height component only. There certainly does appear to be an improvement in the height component accuracy when ambiguities are resolved, almost entirely due to improved (noise-only) precision for such solutions.

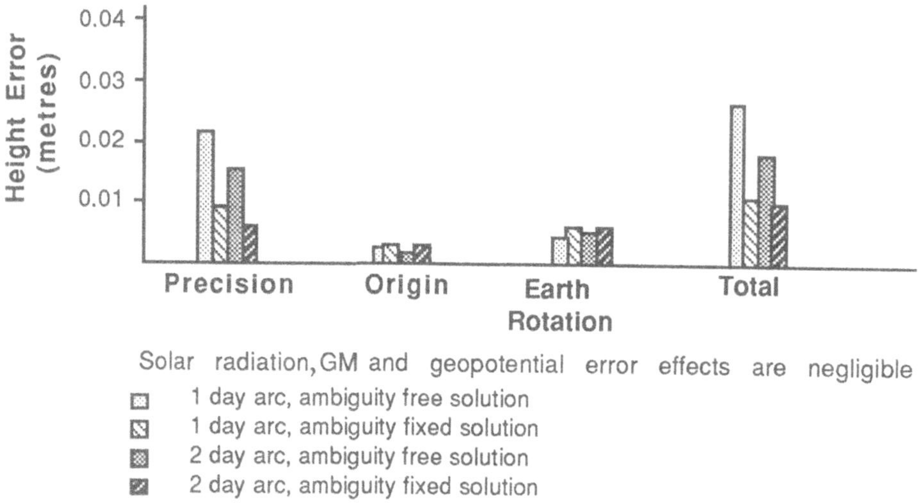

Fig.4. Breakdown of error sources in relative height component of 130km baseline, assuming Free Network option.

The accuracies of the coordinate components of a typical 130km baseline between tide gauges in Bass Strait assuming the Fiducial Network option are shown in Fig. 5, for various arc lengths and for the case where ambiguities are resolved (ambiguity-fixed solution) or left free. A breakdown of the error sources is given in Fig. 6, for the height component only. There is no improvement in the accuracy of the height component when ambiguities are resolved.

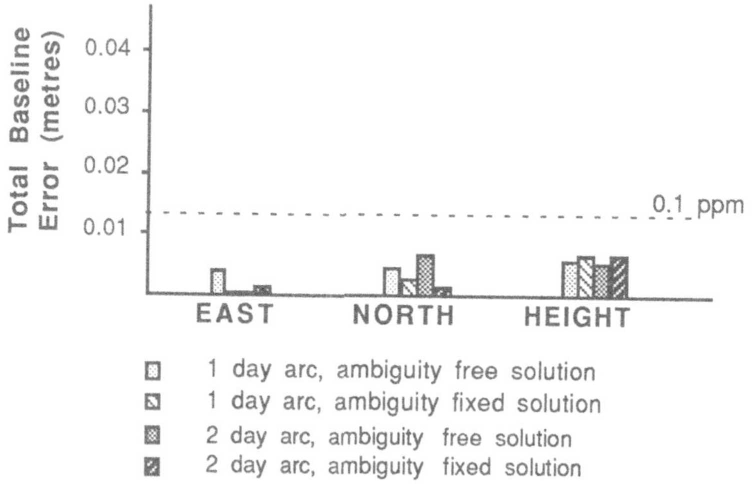

Fig. 5. Accuracy of 130km baseline for 6hr data session(s), assuming Fiducial Network option (Orroral, Yaragadee, Darwin and Wellington sites fixed, Bass Strait sites free).

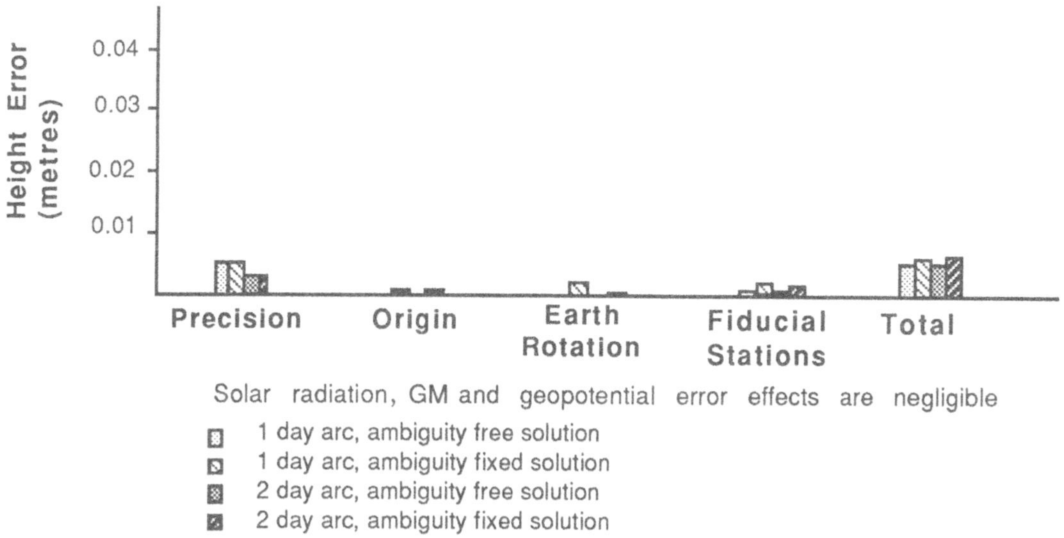

Fig.6. Breakdown of error sources in relative height component of 130km baseline, assuming Fiducial Network option.

Comment. The magnitude of the uncertainty of each individual systematic error source given in Table 1 can be scaled up or down and a new total RSS error obtained. For example, the 0.5m uncertainty in the geocentric coordinates of Orroral may be too pessimistic, and can be changed to, for example, 0.1m, and new RSS values obtained for Figs. 3 to 6.

CONCLUSIONS

It certainly appears feasible that the analysis of data from the Bass Strait GPS campaign will define the relative ellipsoidal heights of tide gauge benchmarks to better than a decimetre. These heights will then serve as the local geometric vertical datum against which the radar altimeter derived sea surface heights (somehow extrapolated into shore) can be compared. The addition of (relative) geoid height information would permit oceanographic data to be referenced to this datum as well.

Acknowledgements. A grant from the Australian Research Council is supporting this project.

REFERENCES

Baines, P.G. and Hubbert, G. (1988). Fluid transport through Bass Strait, proceedings 2nd Conf. on Air-Sea Interaction, Merimbula, N.S.W., 14-17 Feb.

Beutler, G., Gurtner, W., Rothacher, M., Schildknecht, T. and Bauersima, I. (1986). Evaluation of the March 1985 high precision baseline (HPBL) test: fiducial point concept versus free network solutions, *EOS Trans. AGU*, **67**, 911.

Fandry, C.B., Hubbert, G. and McIntosh, P.C. (1985). Comparison of predictions of a numerical model and observations of tides in Bass Strait, *Aust.J.Mar.Freshw.Res.*, **36**,

737-752.
Grant, D.B. (1989). Combination of terrestrial and GPS data for earth deformation studies in New Zealand, UNISURV **S-32**, School of Surveying, University of New South Wales, Australia, in press.
Heiskanen, W. and Moritz, H. (1967). *Physical geodesy*, W.H. Freeman and Co., San Francisco.
Kearsley, A.H.W. (1988). Tests on the recovery of precise geoid height differences from gravimetry, *J. Geophys. Res.*, **93(B6)**, 6559-6570.
Kellogg, J.N., Dixon, T.H. and Neilan, R.E. (1989). CASA Central and South America GPS geodesy, *EOS Trans. AGU*, **70**, 649.
King, R.W., Masters, E.G., Rizos, C., Stolz, A. and Collins, J. (1987). *Surveying with GPS*, Dümmler Verlag, Berlin.
Rizos, C. (1980). The role of the gravity field in sea surface topography studies, UNISURV **S-17**, School of Surveying, University of New South Wales, Australia.
Rizos, C., Govind, R. and Stolz, A. (1989). The Australian GPS orbit determination pilot project: a status report, IAG Symp. Edinburgh, 3-12 August.

THE DETERMINATION OF MEAN SEA LEVEL USING GPS

V Ashkenazi, G A Basker, M Davison & A H Dodson
Institute of Engineering Surveying and Space Geodesy
University of Nottingham, UK

R Hipkin & M Stewart
Department of Geophysics
University of Edinburgh, UK

T Baker
Proudman Oceanographic Laboratory
Bidston, UK

ABSTRACT

In 1986, as a result of the Royal Society's Geodesy Research Initiative, research groups from the University of Nottingham, the University of Edinburgh, and the Proudman Oceanographic Laboratory, in conjunction with the Ordnance Survey of Great Britain, began an investigation into the apparent discrepancy between geodetic and oceanographic methods of levelling. This discrepancy is manifested as a sea slope in a north-south direction around the coast of Great Britain. The project has involved combining GPS measurements with a high precision geoid to provide an independent determination of geoidal height differences on the East coast.

In May 1988 a major GPS campaign was carried out. Measurements were made between the Edinburgh and Lowestoft tide gauges, incorporating other tide gauges en route. Observations were also taken simultaneously at four fiducial stations to allow the computation of precise satellite ephemerides during the period of the campaign.

INTRODUCTION

There are many examples where an anomaly between geodetic and oceanographic levelling has lead to an apparent slope in mean sea level. The single factor common to the apparent sea slopes in Australia, Brazil, Great Britain, Japan and the United States of America is that they occur on a north-south rather than an east-west basis. Subsequent investigations in Australia (Leppert et al, 1975; Angus-Leppan, 1975), California (Sturges, 1974; Balazs et al, 1979), and Great Britain (Edge, 1959; Kelsey, 1970) have failed to satisfactorily determine which of the two levelling methods is in error.

The Second Geodetic Levelling of Great Britain was observed between 1912 and 1938. When it had been completed and adjusted, it was found that there was an apparent difference in mean sea level between Newlyn in Cornwall and Dunbar, near Edinburgh, of 0.81 ft (Edge, 1959). In 1950 an appreciation of the second geodetic levelling indicated that a third geodetic levelling of Great Britain was desirable, using mean sea level observed at Newlyn between 1915 and 1921 as a datum (Ordnance Survey, 1989).

The bulk of the third geodetic levelling was observed between 1952 and 1958, some extra links being added between 1963 and 1968 to tie in a number of tide gauges. The second scientific adjustment of the third geodetic levelling of Great Britain was completed

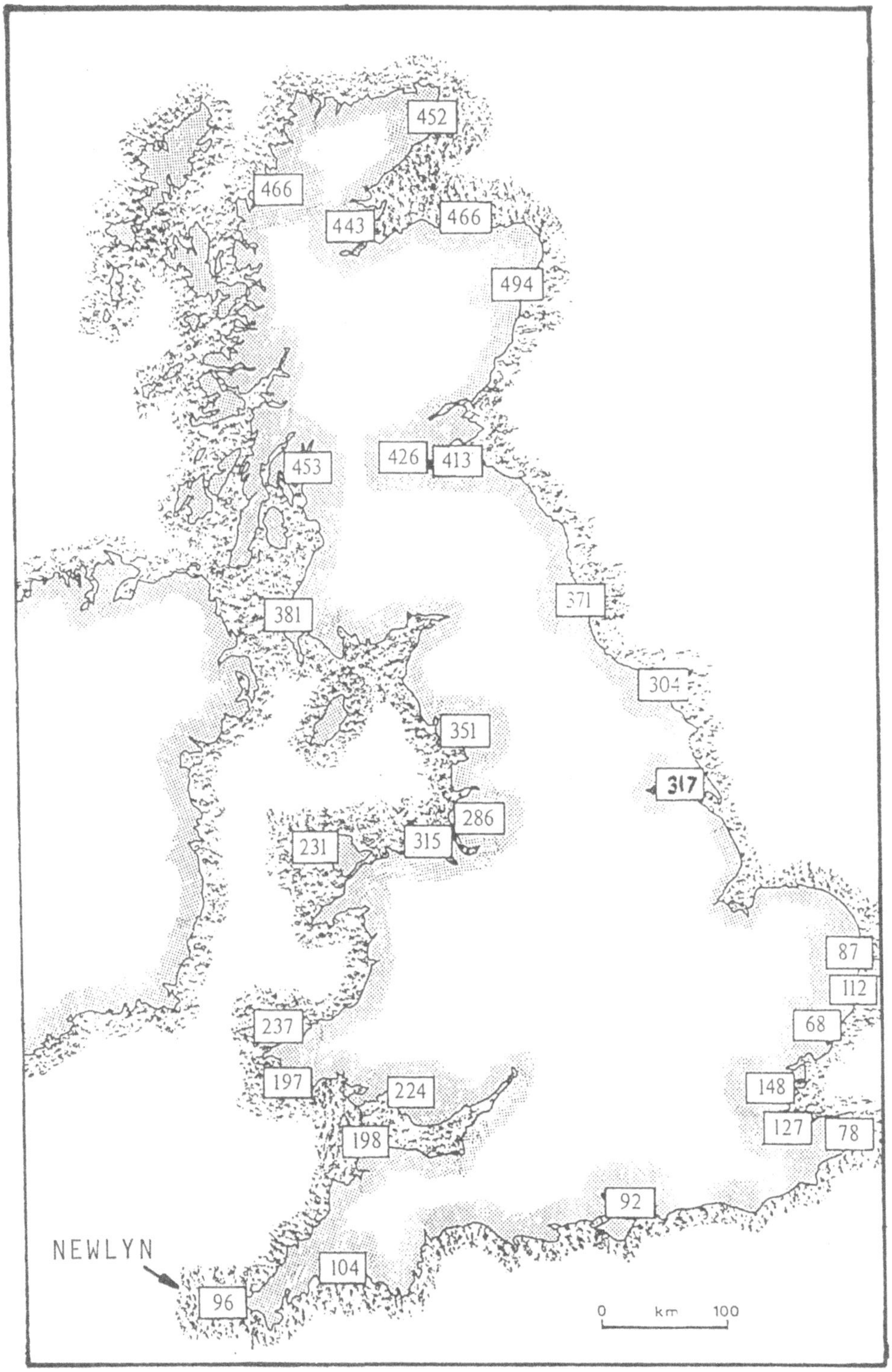

Fig 1 Mean sea level 1960-75 (units are millimetres to Ordnance Datum Newlyn (third Geodetic levelling, second scientific adjustment))

in 1970 (Kelsey, 1970). An analysis of British monthly mean sea levels between 1960 and 1975, observed at the tide gauges and connected to the geodetic levelling (Fig.1), indicated that a latitudinal slope of mean sea level of 5.3 ± 0.4 cm per degree of latitude existed (Thompson, 1980). This northward rise in mean sea level was thought to be difficult to explain oceanographically, and a systematic error in the geodetic levelling was postulated.

There have been three oceanographic determinations of mean sea level using:

1. A monthly regression of monthly mean sea levels against monthly air pressures (Thompson, 1980). This method produces no net slope between Lowestoft and Leith, although the slope varies from +2cm/100km to -2cm.100km.
2. Various numerical hydrodynamic models of the shelf seas and adjacent ocean (Flather, 1981; Davies, 1983; Prandle, 1984). These three methods give similar results, producing a mean slope of approximately -0.4cm/100km between Lowestoft and Leith, again varying between -2.4cm/100km and 1.1cm/100km.
3. A special offshore hydrodynamical levelling exercise along the east coast between Immingham and Leith (Amin, 1988). The slope from Whitby to North Shields was found to be -0.4 ± 0.2cm/100km and from North Shields to Leith the slope was 1.4 ± 0.3cm/100km. The average slope between Leith and Spurn Point was essentially zero.

All three independent and different oceanographic methods give a similar value for the slope of mean sea level which is significantly different to that obtained from the geodetic levelling.

GPS has shown that it is possible to determine accurate ellipsoidal height differences (Engelis et al, 1985). This paper describes the use of GPS with a high precision gravimetric geoid to compute orthometric height differences between the tide gauges at Leith and Lowestoft in an attempt to resolve the discrepancy between geodetic and oceanographic levelling.

The field campaign observed in 1988 will initially be discussed. The processing of the data taken will then be mentioned and some results given. A second field campaign, observed in 1989 will be described, and the paper will be concluded by outlining the remaining work required to complete the investigation. A preliminary account of this project and associated campaigns was given by Ashkenazi et al, 1988.

FIELD CAMPAIGN 1988

This campaign was designed with the intention of solving the previously stated anomaly between the oceanographic and geodetic levelling. Various conclusions had been drawn from a pilot project observed in 1987 (Cleasby, 1987) and from work on a CERI[1] baseline between Jodrell Bank and Defford, and from these following criteria were established for the 1988 field campaign (TGP'88):

1. The use of dual frequency receivers to correct for the first order ionospheric effects.
2. The observations were to be sited, if possible on Ordnance Survey of Great Britain (OS) triangulation pillars.
3. The observing stations were to be sited no further than 30 km apart.
4. Vehicular access to all the observation stations was desirable
5. The observing stations were not to be sited near high tension power cables or radio antennas which might interfere with the GPS data.
6. Observations were to be made at all tide gauges en route.
7. The most southerly baseline observed on one day was to become the most northerly baseline observed on the next. It was hoped that this would provide some check upon the observations between successive days, and also some security in the event that data were lost on one of the days.

---

1. Connected Element Radio Interferometry.

8. Observations were to be made at SLR[2] or VLBI[3] stations throughout Europe so that the so-called `fiducial' concept could be used in the GPS data processing.
9. All possible four satellite constellations were to be observed.

Planning and reconnaissance

TGP'88 may be considered as two separate sections: roving, consisting of those stations required to connect between the tide gauges; and fiducial, well known points occupied for tracking purposes.

Suitable roving stations were chosen by inspecting the relevant maps for the east coast of Great Britain between Leith and Lowestoft (the two tide gauges used as terminals), and then finding the OS triangulation pillars between them. The stations eventually chosen included a number which were sited above rivets due to the lack of an appropriate triangulation pillar, and the maximum separation between two pillars was 33 km (Fig 2).

It was decided to occupy four stations as fiducial points: two of these, Onsala in Sweden and Wettzell in West Germany, are both part of the CIGNET[4] tracking network, and would provide data as a matter of course; the other two fiducial stations were at Jodrell Bank and the Royal Greenwich Observatory at Herstmonceux (RGO) (Fig 3). The latter two required GPS receivers to be provided and manned by personnel from Nottingham.

It was decided to use five roving teams, and seven Texas Instruments TI4100 receivers were consequently hired for the duration of the project.

Observations

The observations were made between 25th May and 1st June 1988. Surface meteorological measurements, of dry and wet temperature and pressure, were taken every fifteen minutes in addition to the GPS data. Three four satellite constellations were observed by the roving receivers on each day and the fiducial stations observed an additional three satellite constellation (Table 1). The data was recorded on IBM PC Convertible micro-computers.

Duration	CH 1	CH 2	CH 3	CH 4
11:30 - 12:20	06	09	11	12
12:50 - 14:00	13	09	11	12
14:00 - 15:10	13	03	11	12
15:10 - 16:10	13	03	--	12

Table 1 TGP'88 satellite observing schedule

DATA PROCESSING

The data has been processed using software developed at the University of Nottingham. After a certain amount of pre-processing has been undertaken, the various data sets from a single session, or combination of sessions, are used as the input to a double differencing network adjustment program called PANIC[5].

2. Satellite Laser Ranging.
3. Very Long Baseline Interferometry.
4. Cooperative International GPS Network.
5. **P**rogram for the **A**djustment of **N**etworks using **I**nterferometric **C**arrier phase observations.

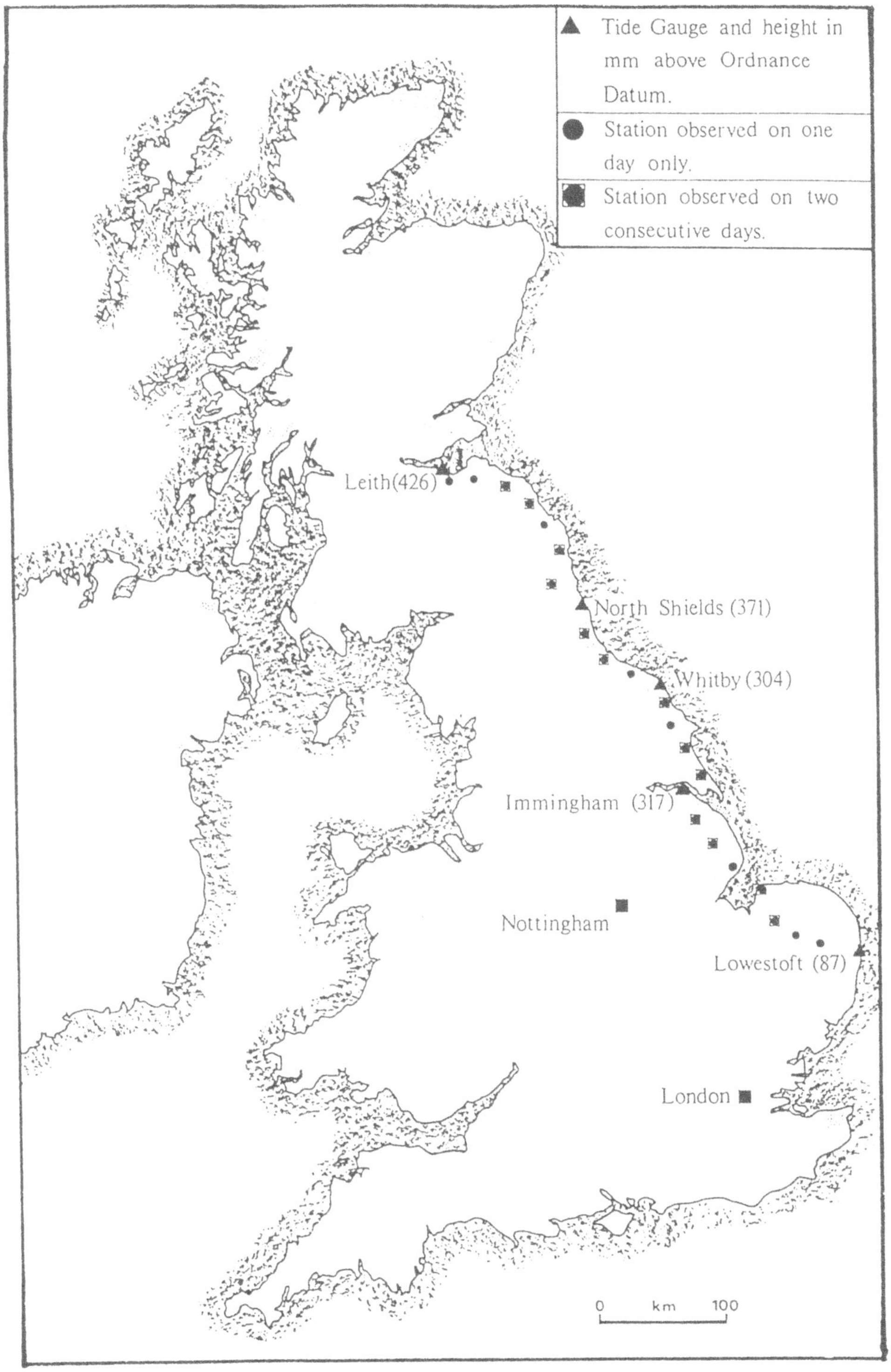

Fig 2 TGP '88 GPS Roving Stations

Fig 3 TGP '88 GPS Fiducial Stations

Not all the fiducial data were compatible with the roving data due to slightly different constellations being observed by the CIGNET stations. There were also data `outages' at the two CIGNET stations. An acceptable solution to this problem was to use the data from the four fiducial stations to determine precise ephemerides for the duration of the project. Two four day solutions were produced.

A number of different strategies have been followed whilst processing the data: two meteorological models, MAGNET (Curley, 1988) and SAASTAMOINEN (Saastamoinen, 1973), have been implemented to correct for the tropospheric delay. The Broadcast ephemeris and precise ephemerides from the Defense Mapping Agency (DMA), University of Texas, and IESSG[6] have been been used. Both L1 and L1/L2 refraction corrected solutions have also been produced.

RESULTS FROM TGP'88

Table 2 shows the results obtained using the various ephemerides. The figures shown are for the overlap baselines observed on successive days.

The Broadcast and DMA ephemerides give a good agreement for the ΔH values. A mean difference, calculated from all the baselines observed gives:

$$\Delta H_B - \Delta H_{DMA} = -0.32 \pm 0.43\text{cm}$$

The Broadcast and IESSG ephemeris give a good agreement in ΔH for the first four day orbit. The agreement for the second four day orbit decrease significantly due to the loss of data at the CIGNET stations. The mean differences for the first and second four day orbits respectively are given below:

$$\Delta H_B - \Delta H_{Fid} = -0.27 \pm 1.08\text{cm}$$

$$\Delta H_B - \Delta H_{Fid} = -3.23 \pm 3.50\text{cm}$$

FIELD CAMPAIGN, JUNE 1989

Some data had been lost at the roving stations in 1988, and this coupled with some of the poor agreements between the overlap baselines observed on successive days lead to the instigation of a limited reobservation campaign in 1989. Three dual frequency Trimble 4000 SLD receivers were used between 10th and 15th July to reobserve a number of the 1988 baselines (Fig 4). Two five satellite constellations were observed, and surface meteorological measurements were taken every thirty minutes. At the time of writing the computation of these reobservations has not been completed.

FUTURE WORK

There is a certain amount of work required to complete the investigation. Firstly, the reobserved baselines have to be computed and the results assessed. The GPS ellipsoidal height differences then have to be converted into orthometric height differences using a precise geoid computed at the University of Edinburgh. This will be followed by computing the height differences between mean sea level at each of the tide gauges, and comparing these values with the corresponding geodetic and oceanographic levelling values.

---

6. Institute of Engineering Surveying and Space Geodesy at the University of Nottingham.

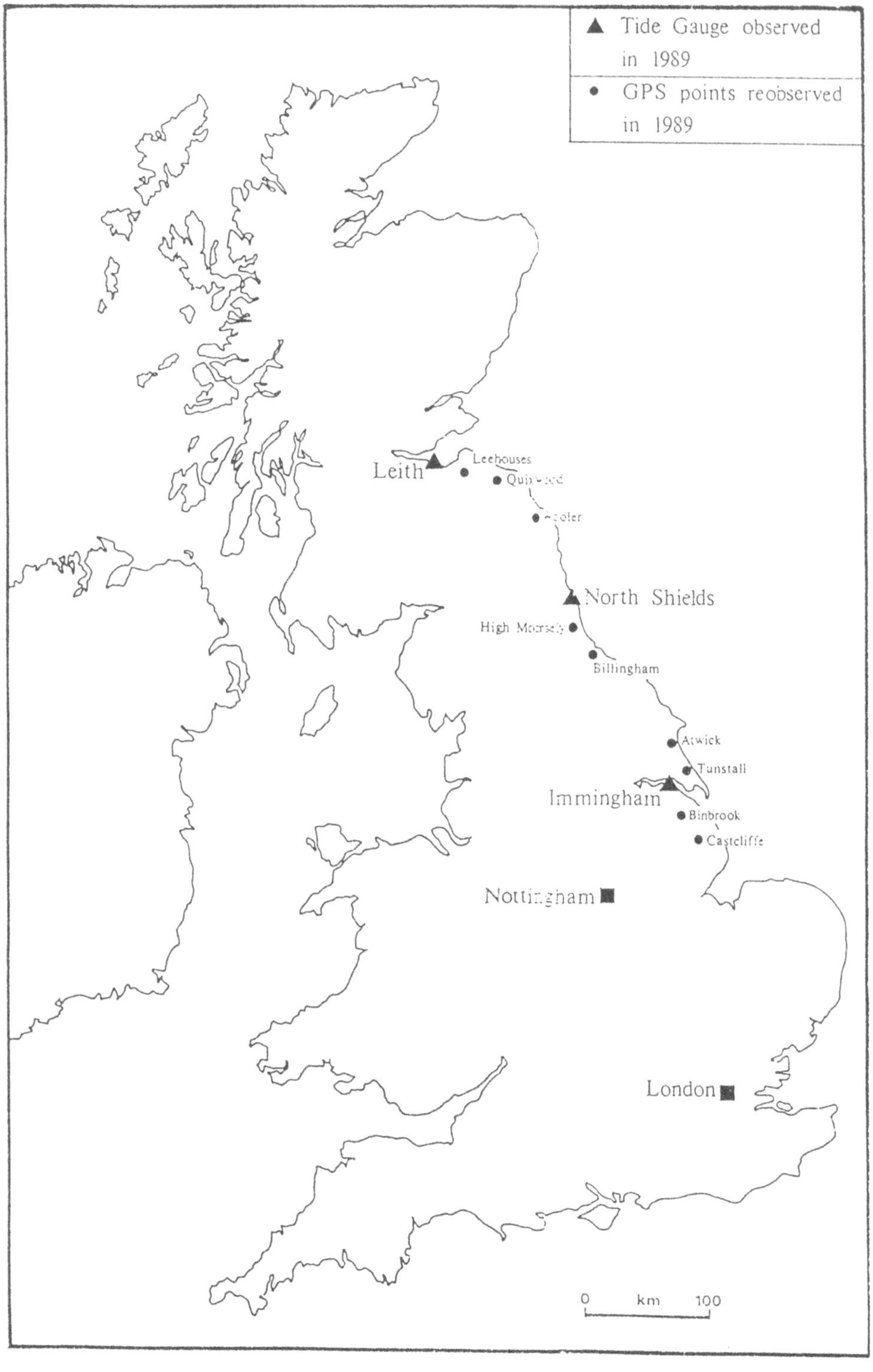

Fig 4 TGP '89 GPS Observation Stations

Site	Baseline (km)	Broadcast (mm)	Ephemeris DMA (mm)	IESSG (mm)
Shi Hel	12.1	26	25	24
Hig Bil	28.9	65	74	84
Whi Clo	17.4	19	17	1
Atw Tun	21.3	69	65	41
Bin Cas	18.7	36	38	2
Tho Gre	23.0	21	19	29

Table 2 A comparison of results obtained using various ephemerides

Acknowledgements

Research on mean sea level determination by GPS at Nottingham University is sponsored by the Natural Environment Research Council, the Science and Engineering Research Council, the Ordnance Survey and the Proudman Oceanographic Laboratory.

REFERENCES

Amin, M. (1988). Spatial variations of mean sea level of the North Sea off the east coast of Britain, *Continental Shelf Res.*, 8(9), 1087-1106, 1988.

Angus-Leppan, P.V. (1975). An investigation of possible systematic errors in levelling along the eastern coast of Australia, *UNISURV G* **23**, 80-92.

Ashkenazi, V., Basker, G.A., Davison, M., Dodson, A.H., Ffoulkes-Jones, G., Moore, T., and Westrop, J. (1988). Levelling of Tide Gauges Using G.P.S. *Presented at American Geophysical Union Chapman Conference on G.P.S. Measurements for Geodynamics, Fort Lauderdale 1988.*

Balazs, E.I. and Douglas, B.C. (1979). Geodetic levelling and the sea slope along the Californian coast, *J. Geophys. Res.* **84**, 6195-6206.

Cleasby, C.V. (1987). The application of GPS in the determination of mean sea level, *MSc Thesis*, University of Nottingham, 1987.

Curley, R.A. (1988). The use of TI4100 receivers and MAGNET software to determine height differences, *MSc Thesis*, University of Nottingham, 1988.

Davies, A.M. (1983). Application of three dimensional shelf model to the calculation of North Sea currents, in *North Sea Dynamics*, (Ed by Sundermann/Lenz), 44-62, Springer-Verlag 1983.

Edge, R.C.A. (1959). Some considerations arising from the results of the second and third geodetic levellings of England and Wales, *Bulletin Géodésique*, No. 52, 28-36, 1959.
Engelis, T., Rapp, R.H., and Bock, Y. (1985). Measuring orthometric height differences with GPS and gravity data, *Manuscripta Geodaetica*, 10:187-194, 1985.
Flather, R.A. (1981). Results from a model of the north-east Atlantic relating to the Norwegian coastal current, in *Proc. of the Norwegian Coastal Current Symposium*, Geilo, 9-12 September 1980, (Ed Saetre and Mork), 427-458, Univ of Bergen, 1981.
Leppert, K., Hamon, B.V., and Mather, R.S. (1975). A status report on investigations of sea surface slope along the eastern coast of Australia, *UNISURV G* **23**, 60 - 67.
Ordnance Survey (1989). *Personal Communication*, 1989.
Prandle, D. (1984). A modelling study of the mixing of ^{137}Cs in the seas of the European continental shelf, *Phil. Trans. Roy. Soc.*, A310, 407-436, 1984.
Saastamoinen, J. (1973). Contribution to the theory of atmospheric refraction II., *Bulletin Géodésique*, No. 107, 13, 1973.
Sturges, W. (1974). Sea level slope along continental boundaries, *J. Geophys. Res.* **79**, 825-830.

RECENT VARIATIONS IN MEAN SEA LEVEL IN SOUTHERN AFRICA

Charles L. Merry
Department of Surveying
University of Cape Town, 7700 Rondebosch, South Africa

INTRODUCTION

The primary reference surface, or vertical datum, for heights is the geoid, which coincides closely with mean sea level. In fact, in many applications the terms are taken as synonymous, although the two may depart from each other by as much as two metres. National or regional vertical datums are usually established by making the implicit assumption that the geoid and mean sea level (MSL) coincide at one or more tide gauge sites, and the tidal records are then used as a basis for heights. As a consequence, any variability in the MSL, temporal or spatial, must be of concern to geodesists and surveyors, as it affects the stability of the vertical datum. Long term trends are also of concern to oceanographers and (taking into account the possible consequences of global warming) also exercise the minds of environmentalists and politicians.

Spatial variations in mean sea level (in practice there is no one single MSL, but only a finite number of measured MSL's) are due to the quasi-stationary departures of sea level from a single equipotential surface - the geoid. As mentioned earlier these departures could reach two metres but are generally less than 50cm. Causes for these departures include the effects of the melting of the polar ice-caps, ocean currents, winds, pressure, and density changes *(Lisitzin, 1974)*. All these effects have time-varying components which cause temporal variations in MSL. The determination of the absolute departure of MSL from the geoid will have to await more precise means of determining absolute geoidal heights, but it is possible to determine differences in the heights of MSL for points which are connected by precise levelling. Such an approach has been used for South Africa *(Merry, 1982)* and for eastern Canada *(Merry and Vaníček, 1983)*, with results consistent with those predicted by oceanographers.

THE DATA

The main emphasis in this paper is on temporal variations in MSL in southern Africa, and on their correlation with pressure variations. Although tidal measurements have been made in South Africa since the late nineteenth century *(Finlay, 1893)* continuous or near-continuous records are only available for the past twenty to thirty years. The operator of the majority of these tide gauges is the Directorate of Hydrography of the South African Navy. Data for this study were obtained from two sources - the Permanent Service for Mean Sea Level (PSMSL), which maintains a global set of monthly mean sea levels, and the Directorate of Hydrography (SADH). The PSMSL data are checked for inconsistencies and reduced to a local reference level before distribution and form the primary data set for this study, with supplementary MSL data being obtained from the SADH. All data are in the form of monthly mean sea levels, and only sites with reasonably continuous records spanning more than nineteen years were considered. As a result of these criteria, no data from east coast sites were used. A summary of the data used is given in Table 1, and the relative positions of the sites are shown in Figure 1. The monthly mean sea level records are shown in graphical form in Figure 2, and some gross errors are already apparent from this figure.

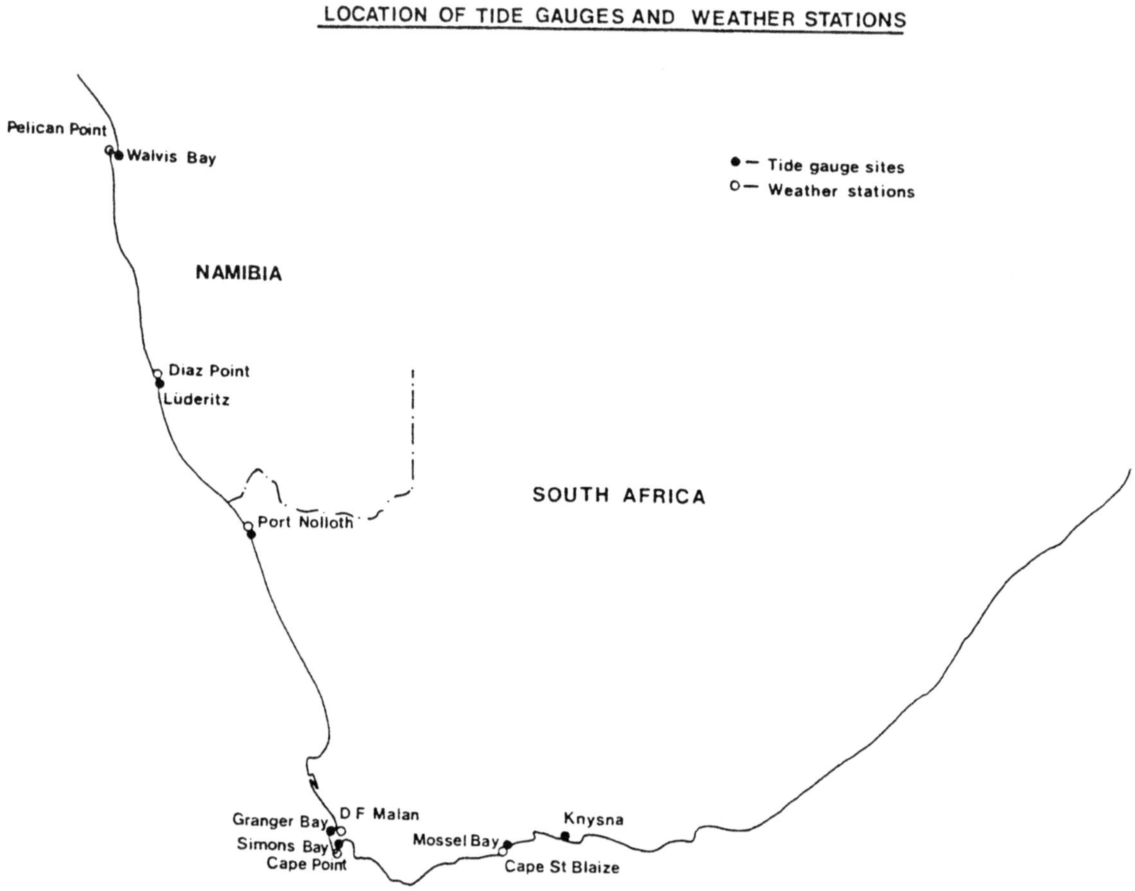

Figure 1: Locality sketch of tide gauge and weather station sites

Table 1: Summary of original MSL data
(mean values refer to an arbitrary local datum)

Site	Time Span	No. of values	Mean (mms)	RMS value (mms)
Walvis Bay	1958-1987	232	-45	68
Lüderitz	1958-1987	329	-76	58
Port Nolloth	1958-1987	311	+59	37
Granger Bay	1967-1987	237	-15	49
Simons Bay	1957-1987	333	+ 6	37
Mossel Bay	1958-1987	291	-37	43
Knysna	1960-1986	299	-18	57

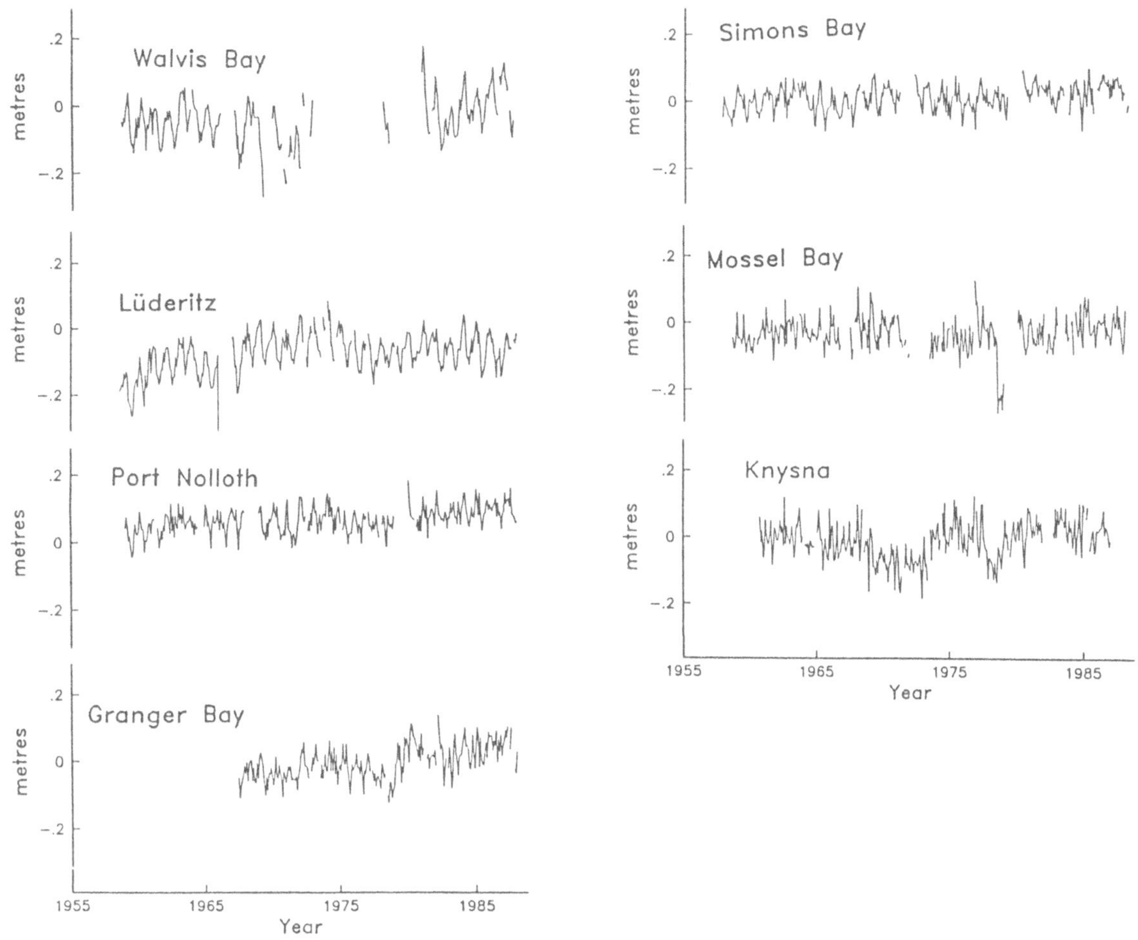

Figure 2 : Monthly Mean Sea Level data

The pressure data (reduced to sea level) were provided by the South African Weather Bureau in the form of thrice-daily readings at weather stations close to the tide gauge sites. In one instance the weather station is located at some considerable distance from the tide gauge site (Cape St. Blaize/Knysna). The positions of these weather stations are also shown on Figure 1. Before the pressure data were used with the MSL data, monthly mean values were computed using simple arithmetic means of the thrice-daily values. The pressure series, corrected for datum biases and linear trends are depicted graphically in Figure 3.

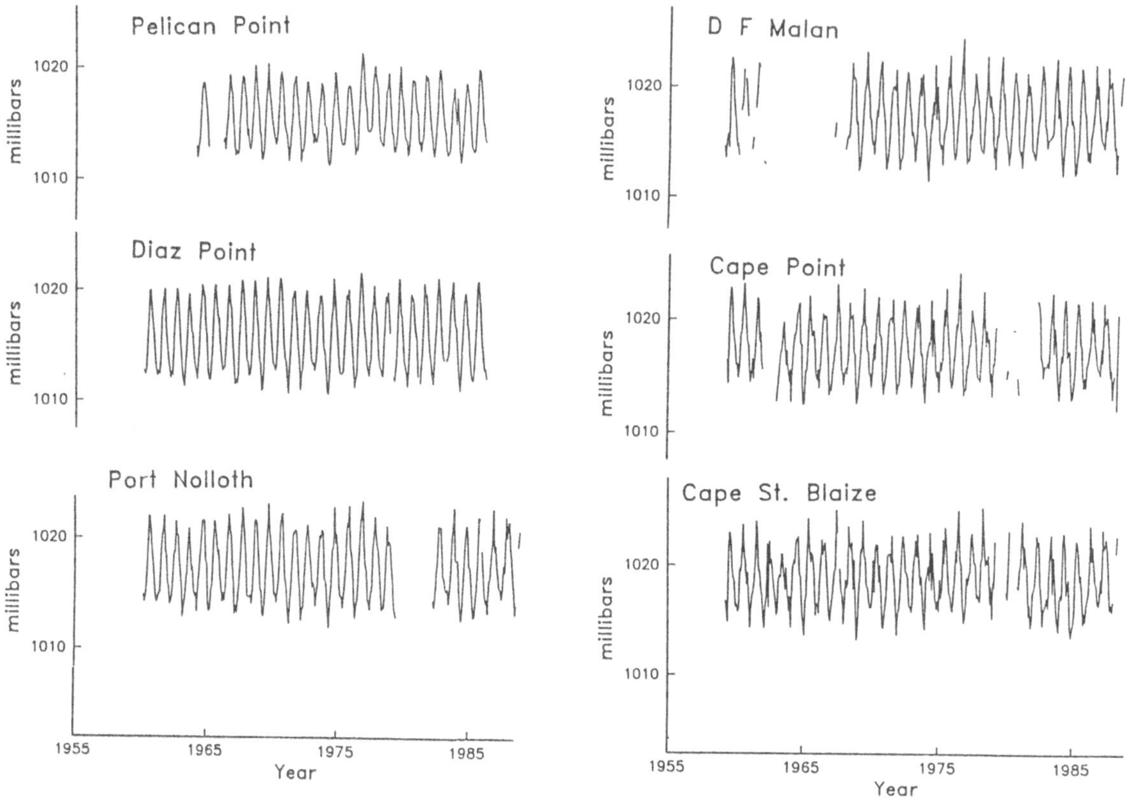

Figure 3 : Corrected pressure data

DATA PROCESSING AND RESULTS

The analysis of the MSL data makes use of the technique of least squares spectral analysis *(Vaníček, 1971)* . This approach has the advantage over conventional Fourier transforms that it can handle gaps in the data and does not require tapering functions to accomodate non-circular functions. As a first step the monthly mean sea levels were subjected to an analysis in which a single datum bias, a linear trend and an annual cycle were determined as part of the process. The results of this analysis are summarised in Table 2, and Figure 4 shows an example of the type of spectrum produced.

Table 2 : Preliminary analysis of MSL data
(phase angle measured relative to 1950.0)

Site	Linear trend (mm per year)	Annual term Amplitude (mms)	Annual term Phase (degrees)
Walvis Bay	+ 2,0 ± 0,4	45 ± 5	48 ± 7
Lüderitz	+ 2,6 ± 0,4	47 ± 3	40 ± 4
Port Nolloth	+ 1,6 ± 0,2	24 ± 2	53 ± 6
Granger Bay	+ 4,1 ± 0,5	24 ± 4	60 ± 8
Simons Bay	+ 1,1 ± 0,2	27 ± 2	36 ± 5
Mossel Bay	0,0 ± 0,2	16 ± 3	32 ± 12
Knysna	+ 1,1 ± 0,5	1 ± 4	98 ± 294

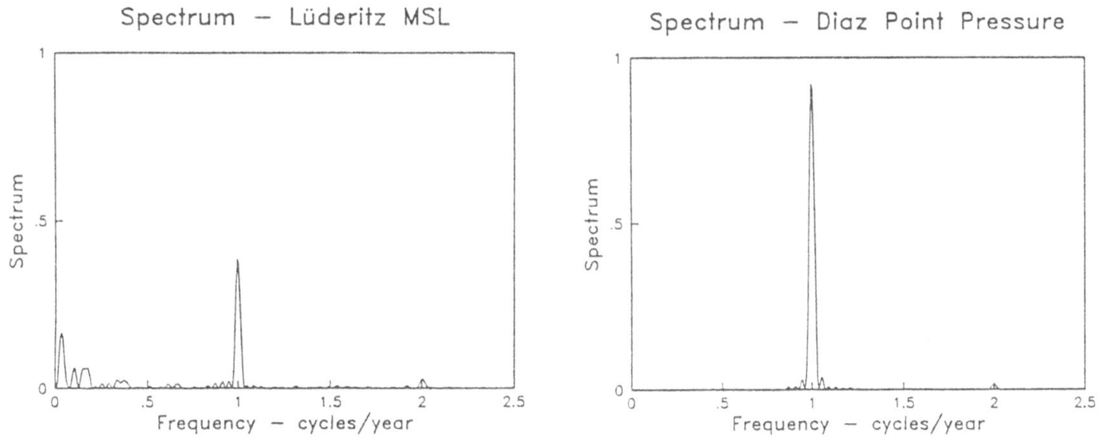

Figure 4 : Spectra for Lüderitz (MSL) and Diaz Point (pressure)

This first analysis enabled data outliers and predominant frequencies in the data to be identified. In a few cases visual inspection of graphs of the residuals enabled possible datum biases to be identified. Datum biases (other than the initial bias) are very worrying in this type of record. If small, they are difficult to isolate against the background noise. Furthermore, they are highly correlated with linear trends and it is often impossible to determine whether a tendency in the time series is due to a datum bias or a genuine trend in MSL. Where one has access to the original field records it is often possible to resolve this problem, as these notes may indicate a breakdown or movement of the gauge. Another indicator of a possible datum shift is a short steep trend in the record, followed by a data gap. Although all tide gauges are connected to nearby benchmarks, there is no regular programme of relevelling, and hence no opportunity to detect local vertical movements of the jetty. In our case we only had access to "cleaned" data and could not confirm any of the suspected datum biases (attempting to solve for these biases as part of the least squares analysis resulted in shifts which were insignificantly different from zero at all sites where this was attempted). On the other hand, individual outliers could be clearly identified at Lüderitz and Mossel Bay (see Figure 2) and were removed.

The pressure time series needed some refinement before they could be used to reduce the variance in the MSL series. Visual inspection and least squares analysis of the pressure data revealed some large inconsistencies which would have biased the MSL data. For example, at Cape Point a discontinuity of 26,8 mbars occurs at the start of 1978. Similar steps, of smaller magnitude, occur for most of the other weather stations at the same time. The source of these steps is still being sought but is believed to be related to a change of instrumentation. After removal of these and other smaller biases, further analysis of the pressure records revealed a consistent decrease in pressure with time at all sites, ranging from -0.02mbar per year to -0.07mbar per year. Although small and barely significant, the use of these unaltered pressure series to reduce the MSL records could have introduced a false trend in MSL of as much as 0,7mm per year. To avoid this the linear trends were removed from the pressure data. Again, the source of these trends in the pressure data is unknown and requires further investigation. The pressure records shown in Figure 3 represent these corrected data.

The predominant periodic term in the pressure and MSL series is the annual cycle, as evidenced by the spectral analysis (Figure 4). The root mean square (RMS) variations in the MSL series can be reduced by subtracting the effect of the pressure variations, thus providing a less noisy series from which more reliable estimates may be obtained for the linear trend in sea level. The theoretical "inverted barometer" relationship between pressure and sea level contains an amplitude ratio of 10mm per mbar, with a phase difference of 180°. However the relationship is complicated by the impact of other annual phenomena - some, such as wind setup, directly related to air pressure, others, such as river discharge, not. Empirical investigations *(Merry and Vaníček, 1983; Pugh, 1987)* have shown that the actual relationship can differ significantly from that indicated by theory. The least squares response method *(Steeves, 1981)* provides a convenient means of deriving the amplitude and phase response of sea level to atmospheric pressure variations. We have applied this solely to the annual period, with the results as shown in Table 3. With the exception of Knysna the responses do not differ markedly from that expected from theory. In the case of Knysna the small response and non-180° phase lag could be due to a combination of the gauge being sited on a lagoon with a narrow entrance and out-of-phase stream inflow into the lagoon.

Table 3 : Response analysis - MSL/Pressure
(annual cycle only; phase lag taken in the sense: $\phi_{press} - \phi_{MSL}$)

Sites	Amplitude ratio (mm/mbar)	Phase lag (degrees)	RMS value of reduced MSL (mms)	Percentage reduction in RMS
Walvis Bay	13,6	178	67	2
Lüderitz	11,8	180	40	31
Port Nolloth	6,3	169	30	23
Granger Bay	6,3	162	43	12
Simons Bay	8,4	190	27	27
Mossel Bay	5,5	191	36	16
Knysna	0,3	125	55	4

An average response of 8mm per mbar and a phase difference of 180° were used to remove the influence of the atmospheric pressure from the tidal records. No attempt was made to interpolate gaps in either the MSL or pressure records with the result that the reduced MSL records exhibit a much more intermittent pattern than the original records (see Figure 5). This is especially the case for Walvis Bay, where the pressure observation epochs seldom coincided with the MSL epochs. The removal of the pressure signal has reduced the RMS variations of all the MSL series, although not by as much as had been expected. The small reduction at Walvis Bay is no doubt due to the intermittent pressure and MSL records, while that at Knysna is due to the lack of coherence of the two series. Discounting these two sites, the results are in good agreement with those obtained by *Brundrit (1984)*, using shorter records.

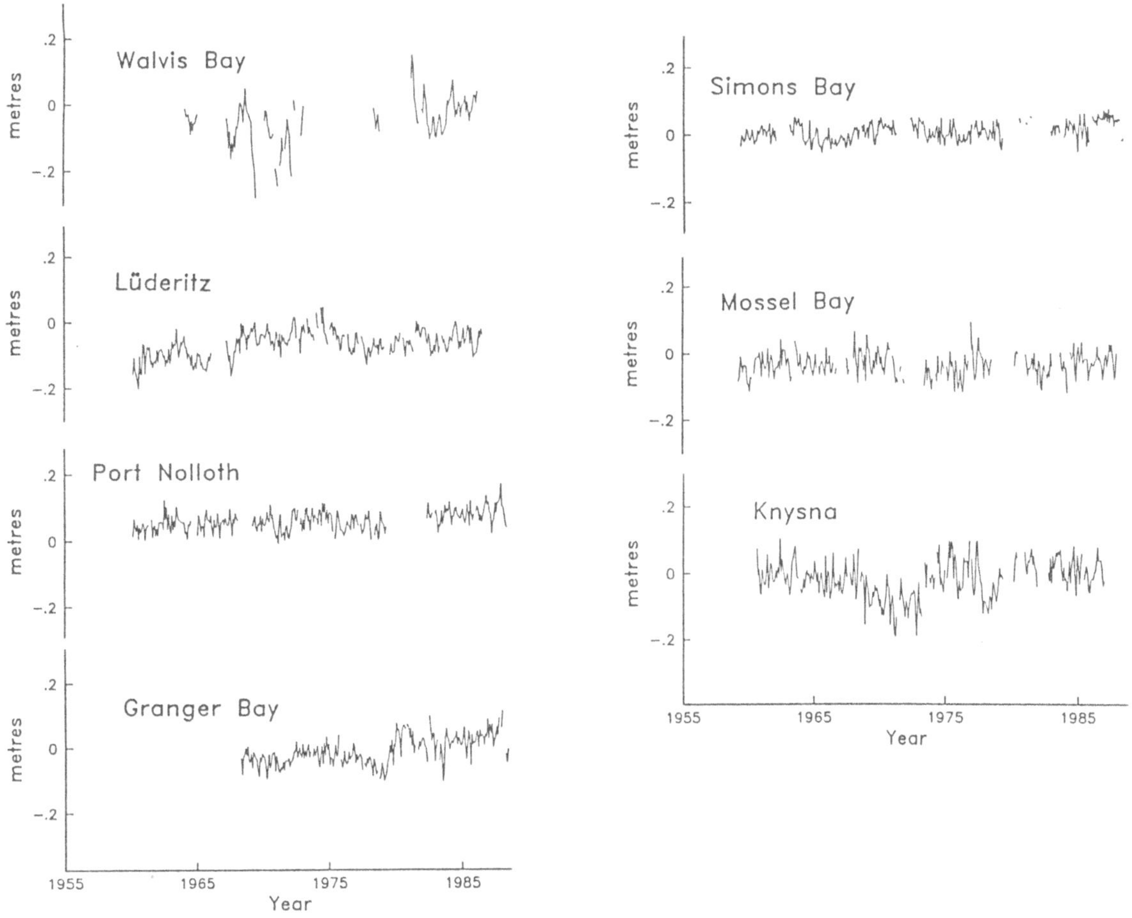

Figure 5 : Reduced Mean Sea Level data

A further spectral analysis of these reduced MSL records has been carried out, together with a simultaneous search for linear trends and for any residual annual terms. The results of this analysis are summarised in Table 4. The effect of using a single response factor for the effect of atmospheric pressure is evident in this table. The residual annual terms are not insignificant and their source (e.g. wind set-up) needs further investigation. Comparison of Tables 2 and 4 show that, with one exception, little change has occurred in the resolved linear trends after removal of the pressure signal. The exception is Walvis Bay, where a lot of data have been lost due to gaps in the pressure record, and the residual MSL series differs markedly from the original. Even then the linear trend retains the same sign and is not wildly dissimilar in magnitude.

With the exception of Mossel Bay (no significant trend) all the analyses show a significant rise in sea level, averaging 1,9mm per year over the past 20 to 25 years. This trend is generally higher on the west coast than on the south coast. Visual inspection of some of the MSL records (e.g. those for Port Nolloth and Granger Bay) indicates a possible increase in this rise over the past ten or so years. However the data are certainly not of sufficient quality to confirm this last suspicion.

Table 4 : Analysis of MSL data
(phase angle measured relative to 1950.0)

Site	Linear trend (mm per year)	Annual term Amplitude (mms)	Phase (degrees)
Walvis Bay	+ 3,5 ± 0,7	17 ± 8	46 ± 25
Lüderitz	+ 2,2 ± 0,2	18 ± 3	31 ± 9
Port Nolloth	+ 1,4 ± 0,2	8 ± 2	182 ± 16
Granger Bay	+ 4,0 ± 0,5	11 ± 3	174 ± 17
Simons Bay	+ 1,1 ± 0,2	4 ± 2	36 ± 34
Mossel Bay	0,0 ± 0,2	11 ± 3	235 ± 15
Knysna	+ 1,0 ± 0,4	27 ± 4	216 ± 9

The determined trends are consistent with those determined elswhere in the world *(Pugh, 1987)*, both in sign and in magnitude. Although several commentators have predicted a quickening in the rise in global sea level due to the greenhouse effect, this can not yet be confirmed in the southern African records. How does all this effect the local geodesist and surveyor? It was already shown some eight years ago *(Merry, 1982)* that the current levelling datum in South Africa is some 15 to 20cm below MSL. This difference must already be taken into account in constructing harbour works. As time progresses this discrepancy will increase and will need to be taken into account when recomputation of the South African vertical networks takes place. The time-varying aspect of MSL will also need to be considered, and it is possible that an alternative definition of the vertical datum, free of the vagaries of sea level could be selected. Discussion of this topic, however, is beyond the scope of the present paper - for a review see, for example, *Vaníček et. al., 1987*. What is important is that a trend in sea level exists and that this trend varies significantly from one site to another - further complicating the use of mean sea level as a basis for vertical datum definition in southern Africa.

CONCLUSIONS

The currently available MSL data around southern Africa's coast are noisy and intermittent. Careful analysis is needed to reduce this noise and to detect possible linear trends resulting from a possible global warming. Even then the possibility exists that undetected changes in datum may bias these determinations. Nevertheless it is encouraging that almost all the determined trends are of one sign and similar magnitude, and are consistent with those determined elsewhere in the world. No increase in this trend can be confirmed, although the series cover too short a time span and are too noisy for either confirmation or denial of such a possibility.

Some enhancement of the procedures used here is possible. Interpolation of the pressure records could be carried out, provided it does not bias the results. The effects of winds on the MSL records could be investigated, although this may prove difficult due to the paucity of wind data near the tide gauge sites. This problem could be resolved by using the pressure data to compute the geostrophic wind, but this approach is hampered by the need for several pressure sites in the vicinity of each tide gauge. Nevertheless the attempt should and will be made.

Acknowledgements: Data for this analysis were made available by the Permanent Service for Mean Sea Level, the South African Directorate of Hydrography and the South African Weather Bureau and their contributions are gratefully acknowledged. The software for the least squares spectral analysis was provided by Prof. P Vaníček of the University of New Brunswick. Financial support for this research was provided by the Foundation for Research Development and by the University of Cape Town.

REFERENCES

Brundrit, G.B. (1984). Monthly mean sea level variability along the west coast of southern Africa. *South African Journal of Marine Science,* **2**.

Finlay, W.H. (1893). Approximate tidal constants for Table Bay and Algoa Bay. *Transactions, South African Philosophical Society,* **5**.

Lisitzin, E. (1974). *Sea level changes.* Elsevier, Amsterdam.

Merry, C.L. (1982). Variations in mean sea level on the South African coast. *Proceedings, Seventh Conference of Southern African Surveyors,* Johannesburg.

Merry, C.L. and P. Vaníček (1983). Investigations of local variations of sea-surface topography. *Marine Geodesy,* **7**(1-4).

Pugh, D.T. (1987). *Tides, surges and mean sea level.* John Wiley and Sons, Chichester.

Steeves, R.R. (1981). Least squares response analysis and its application to tiltmeter observations. Unpublished PhD thesis, University of New Brunswick, Fredericton.

Vaníček, P. (1971). Further development and properties of the spectral analysis by least squares. *Astrophysics and Space Science,* **12**.

Vaníček, P., P.A. Cross, J. Hannah, L. Hradilek, R. Kelm, J. Maakinen, C.L. Merry, L.E. Sjöberg, R.R. Steeves, D.B. Zilkowski (1987). Four dimensional geodetic positioning. *Manuscripta Geodaetica,* **12**(3).

A GENERAL OCEAN CIRCULATION MODEL DETERMINED IN A SIMULTANEOUS SOLUTION WITH THE EARTH'S GRAVITY FIELD

R. Steven Nerem †, Byron D. Tapley, and C. K. Shum
Center for Space Research, The University of Texas at Austin
Austin, Texas 78712 U.S.A.

INTRODUCTION

Over the past several years, determination of the sea surface topography using satellite altimetry has been greatly improved due to the implementation of a technique which simultaneously solves for the sea surface topography, the gravity field of the Earth, and the satellite orbit [*Tapley et al.*, 1988; *Marsh et al.*, 1989; *Nerem*, 1989; *Engelis and Knudsen*; 1989]. The sea surface topography, ζ, is defined as the deviation of the ocean surface from the geoid due to the geostrophic component of the ocean currents. If ζ is known, then the surface geostrophic velocity of the ocean currents may be computed using basic hydrodynamic equations as [*Coleman*, 1980]:

$$u = -\frac{g}{f r_e}\frac{\partial\zeta}{\partial\phi} \qquad v = \frac{g}{f r_e \cos\phi}\frac{\partial\zeta}{\partial\lambda} \tag{1}$$

where u and v are the velocity components in the longitude (λ) and latitude (ϕ) directions respectively, g is the acceleration of gravity, $f = 2\Omega\sin\phi$ is the Coriolis force coefficient, Ω is the rotation rate of the Earth, and r_e is the mean equatorial radius of the Earth. When used in combination with appropriate measurements of the ocean density versus depth, it is possible to model the complete three dimensional circulation of the ocean [*Wunsch and Gaposchkin*, 1980]. This may be one of the more important results future altimeter missions will provide.

Satellite altimetry has been used in the past to determine models of ζ [*Douglas et al.*, 1984; *Tai and Wunsch*, 1984; *Engelis*, 1984], but the degree of success has been limited by the errors in the model of the Earth's gravity field, which effects both the determination of the geoid and the estimation of the satellite orbit. These early investigations analyzed the difference between a mean sea surface (computed with SEASAT altimeter data and an independent estimate of the satellite orbit) and an independently computed long wavelength gravity field model. These differences were filtered in order to extract the long wavelength component of ζ from the shorter wavelength geoid errors. The fact that the gravity field, the satellite orbit, and the model of ζ were all computed independently prevented the information in the altimeter data from contributing to the solution for the gravity field or the satellite orbit. Solving for these three parameter sets simultaneously allows the information

† **Now with the University of Maryland Astronomy Program at NASA/Goddard Space Flight Center, Code 626, Space Geodesy Branch, Greenbelt, Maryland, 20771, U.S.A.**

in the altimeter data to be separated into orbit information, gravity model information, and information on ζ. This results in improved models for each of these phenomena. The method employed in this investigation computes the models of the satellite orbit, the gravity field of the Earth, and ζ in a simultaneous dynamically consistent solution using GEOSAT altimeter data [*McConathy and Kilgus*, 1987] and doppler tracking data.

METHODS

The data employed in this investigation consists of laser range and doppler tracking data from 14 different satellites including GEOSAT, altimeter crossover data from GEOSAT, direct GEOSAT altimeter data processed in the form of altimeter normal points, and surface gravity data. The satellites in addition to GEOSAT whose tracking data were employed in the solution include LAGEOS, Ajisai, Starlette, NOVA-1, Oscar-14, GEOS-1, GEOS-2, GEOS-3, BE-C, DI-D, DI-C, Peole, and SEASAT. The surface gravity data was supplied in the form of normal equations computed from 1° x 1° mean terrestrial gravity anomalies by Ohio State University [*Pavlis*, 1988].

51 days of GEOSAT altimeter data and doppler tracking data in the form of three 17 day arcs (11/17/86-12/4/86, 12/21/86-1/24/87) were included in the solution [*Smith*, 1988]. The GEOSAT altimeter data was received in the form of Geophysical Data Records spaced at 1 second intervals. This data was then preprocessed to form altimeter normal points with a 10 second data spacing. This was accomplished by first removing the high degree terms (greater than degree and order 50) of the geoid from the altimeter data using a model for the short wavelength geoid and then fitting a straight line to altimeter data in 10 second windows. The altimeter normal points were created by evaluating the height of the straight line at its midpoint for each 10 second window. The high degree geoid correction was necessary to prevent aliasing since only a long wavelength gravity model was estimated. Altimeter normal points were computed to compress the altimeter data and conserve computer time.

The Earth's gravitational potential, the gradient of which is used to compute the gravitational force acting on the satellite, is modeled using a spherical harmonic expansion as:

$$U = \frac{\mu}{r}\left[1 + \sum_{l=2}^{L_{max}} \sum_{m=0}^{l} \left[\frac{r_e}{r}\right]^l \overline{P}_{lm}(\sin\phi)\left[\overline{C}_{lm}\cos m\lambda + \overline{S}_{lm}\sin m\lambda\right]\right] \qquad (2)$$

where $\overline{C}_{lm}, \overline{S}_{lm}$ are the normalized spherical harmonic geopotential coefficients of degree l and order m, μ is the product of the gravitational constant and the total mass of the Earth, r_e is the mean equatorial radius of the Earth, r,ϕ,λ are the radial distance, geocentric latitude and longitude measured from the zero meridian, $\overline{P}_{lm}(\sin\phi)$ is the normalized associated Legendre function of degree l and order m ($m \neq 0$) or the normalized Legendre polynomial of degree l ($m=0$), and L_{max} is the truncation degree of the geopotential ($L_{max}=50$ for the solution discussed here). The $\overline{C}_{2,0}$ coefficient is modeled to include the permanent deformation of the Earth due to the Sun and the Moon. The position vector of the sub-satellite point on the geoid (x,y,z) is computed by iteratively solving the following two equations [*Shum*, 1983]:

$$W(x,y,z) = W_0 \quad \text{and} \quad \nabla W / \left|\nabla W\right| = \hat{u}_h \qquad (3)$$

where $W(x,y,z) = U + T$, the total potential at the sub-satellite point, T is the rotational potential, W_0 is the value of the total potential corresponding to the mean sea surface, and

$\hat{u}_h$ is a unit vector normal to the geoid and passing through the satellite. Thus the geopotential coefficients, $\overline{C}_{lm},\overline{S}_{lm}$, are used both to compute the satellite dynamics and the geoid height. This coupling allows the solution to discriminate gravity-induced orbit errors from geoid errors. The height of ζ is modeled using surface spherical harmonics as:

$$\zeta(\phi,\lambda) = \sum_{l=1}^{l_{max}} \sum_{m=0}^{l} \overline{P}_{lm}(\sin\phi)\left[\overline{C}'_{lm}\cos m\lambda + \overline{S}'_{lm}\sin m\lambda\right] \tag{4}$$

where $\overline{C}'_{lm},\overline{S}'_{lm}$ are the normalized spherical harmonic coefficients of the model for ζ.

A linear modified least squares estimation technique was employed to determine the spherical harmonic coefficients of the geopotential and ζ, and the parameters modeling the orbit simultaneously. An *a priori* constraint was applied to the geopotential coefficients and the coefficients of ζ. The *a priori* information consisted of an *a priori* value of zero and an associated variance for each coefficient (both the geopotential and ζ). The variance for the geopotential was specified by a modified form of Kaula's Rule [*Kaula*, 1966; *Wagner and Colombo*, 1979] as:

$$\sigma_l^2 = \frac{1}{2}\frac{10^{-10}}{l^4} \tag{5}$$

The *a priori* variance of ζ was specified by a power rule derived from an oceanographic model of ζ [*Levitus*, 1982; *Nerem*, 1989] as:

$$\sigma_l^2 = 0.0361 \cdot l^{-2.54} \tag{6}$$

The use of the *a priori* variance for the geopotential causes poorly observed coefficients to tend toward zero with an error variance equal to the *a priori* variance. With the *a priori* information employed for ζ: 1) the value of ζ over land areas tends towards zero with an error variance equal to the *a priori* variance, and 2) the high degree coefficients of ζ (degree 10) which are difficult to separate from the gravity field will tend toward zero and all the signal will be put into the gravity solution.

The parameters estimated in the simultaneous solution are the following:

- Spherical harmonic coefficients of the geopotential complete to degree 50.
- Spherical harmonic coefficients of ζ complete to degree 10.
- Satellite position and velocity at epoch for each arc in the solution.
- Coefficients for atmospheric drag, solar radiation pressure, etc.
- Ocean tide parameters (satellite dynamical effects only).
- Selected tracking station coordinates (including GEOSAT doppler stations).
- GEOSAT altimeter bias and scale factor to $H_{1/3}$ for the EM bias correction.

After the spherical harmonic coefficients of ζ were obtained, the solution was corrected *a posteori* so that the model would be referenced as closely as possible to an equipotential surface. The second degree zonal coefficient of ζ was corrected by adding 0.0885 to the nominal value in order to account for the potentials of the sun and the moon. Thus the resulting model of ζ is referenced to an equipotential surface defined by the potentials of the Earth, Sun, and Moon collectively rather than the Earth alone.

RESULTS

Figure 1 shows a contour map of ζ computed using Equation (4) and the spherical harmonic coefficients of ζ determined from GEOSAT altimeter data in a simultaneous solution for the spherical harmonic coefficients of the gravity field and the satellite orbit parameters. The model is complete to degree and order 10 in harmonics. For comparison, Figure 2 shows a contour map of ζ determined by fitting a spherical harmonic model to a gridded 1° x 1° model of ζ computed from a global data base of oceanographic data collected by ships at sea [*Levitus*, 1982]. Levitus compiled an archive of hydrographic data collected over a 70 year time period ending in 1977. This data base was then used to compute the height of ζ in 1° x 1° bins using the 2250 dbar pressure surface as a reference level. A spherical harmonic model complete to degree and order 10 was then fit to the gridded model using a least squares procedure.

The two maps compare quite well in most areas of the world with surprisingly small differences. The circulation features in the North and South Atlantic and the Indian Ocean are better defined in the altimeter model of ζ than in the hydrographic model of ζ. This may be caused by the significant increase in temporal averaging for the hydrographic results (70 years) as compared with the altimeter results (3 months). The representations of the circulation features in the Pacific Ocean are comparable, although the locations of the centers of the features are shifted northeastward slightly in the altimeter results relative to the hydrographic results. The Antarctic Circumpolar Current is somewhat better represented in the hydrographic results, primarily because the altimeter data used to compute the GEOSAT results extends to only 60° S latitude. In addition, the data base used by Levitus to construct his map had very few measurements in the southern hemisphere [*Levitus*, 1982], thus differences with the altimeter results in this area would not be surprising. The use of the 2250 dbar reference level in the hydrographic computations also causes differences unless this level corresponds to an equipotential surface globally, which is unlikely.

The gravity model solution can be used with the doppler tracking data to compute new GEOSAT orbit estimates. These orbit estimates can be used along with the improved models for the geoid and ζ to compute altimeter residuals (the misclosure between the model of the altimeter measurement and the actual measurement). For the 51 days of GEOSAT data used in this investigation, the altimeter residuals have a global root-mean-square (rms) value of 45 cm, although large areas of the ocean have altimeter residuals with an rms value of 25 cm. The altimeter residuals are largest in areas with high ocean current variability or in areas where the ocean tide models are inaccurate. The same is also true of the rms of the altimeter crossover residuals, which have a global rms value of 22 cm, but are less than 15 cm over large areas of the ocean surface.

ERROR ANALYSIS

The errors in the simultaneous solution may be analyzed using the error covariance matrix of the solution. The error covariance matrix contains the error variances and correlations for the spherical harmonic coefficients of the geopotential and ζ as well as other parameters in the solution. The portion of the covariance matrix corresponding to the spherical harmonic coefficients of the geopotential and ζ may be linearly mapped to determine the estimated errors in the geoid height, the errors in ζ, and the correlation of the errors as a function of latitude and longitude.

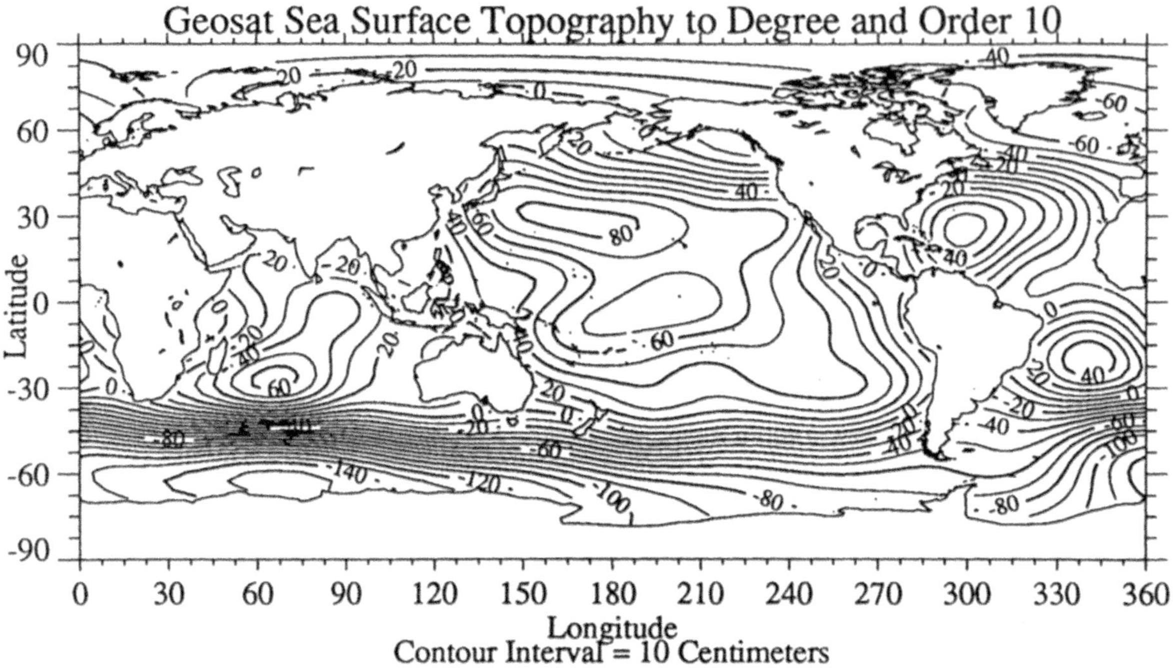

Fig. 1. Map of the ζ complete to degree 10 derived from GEOSAT altimeter data.

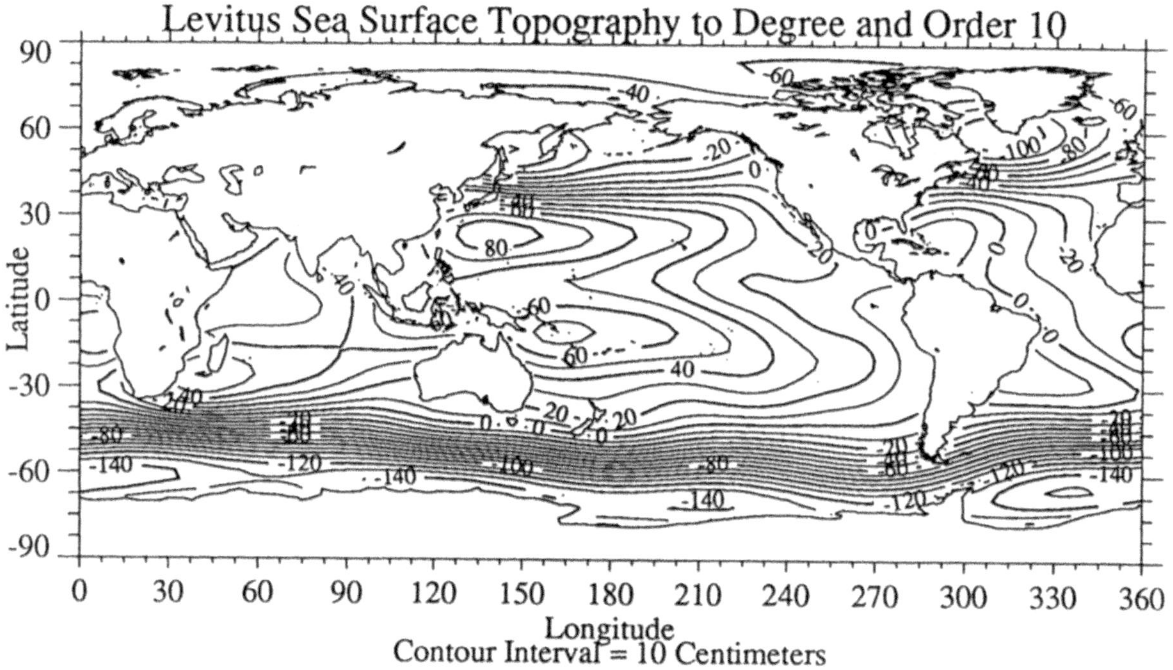

Fig. 2. Map of ζ complete to degree 10 derived from hydrographic data [*Levitus*, 1982]

Figure 3 is a map of the estimated comission errors of the geoid for a model complete to degree and order 50. These errors average 35 cm over the oceans which is consistent with the altimeter residuals reported in the previous section. This is a substantial improvement (35 cm versus 1.5 meters) over previous solutions which did not contain altimeter data. Figure 4 is a map of the estimated errors in ζ for a model complete to degree and order 10. These errors average 13 cm in the open ocean, but are slightly larger near the land areas. This effect was most likely caused by the editing of altimeter data in water shallower than 2000 meters. The correlation coefficient of the errors in the geoid and ζ, $\rho_{N,\zeta}(\phi,\lambda)$,may be computed as:

$$\rho_{N,\zeta}(\phi,\lambda) = \frac{C_{N,\zeta}(\phi,\lambda)}{\sigma_N(\phi,\lambda)\cdot\sigma_\zeta(\phi,\lambda)} \tag{7}$$

where $\sigma_N(\phi,\lambda)$ is the standard deviation of the geoid height, $\sigma_\zeta(\phi,\lambda)$ is the standard deviation of ζ, and $C_{N,\zeta}(\phi,\lambda)$ is their cross covariance. The correlation coefficient has a value of ±1 if the errors are perfectly correlated (thus indicating that the chosen parameters are inseparable) and a value of 0 if the errors are completely uncorrelated. Generally, correlation coefficients less than .5 are not considered significant. Over most of the ocean, the errors of the geoid and ζ have a correlation coefficient with a magnitude less than 0.2, indicating that excellent separation of the parameters has been obtained in the solution.

By comparing the errors in the geoid and the signal of ζ by degree, the maximum degree to which the model of ζ may be separated from the geoid can be estimated. The degree error variance of the geoid height, σ_l^2, and the degree variance of ζ, d_l^2, may be computed as:

$$\sigma_l^2 = r_e^2 \sum_{m=0}^{l} \left(\sigma_{C_{lm}}^2 + \sigma_{S_{lm}}^2\right) \qquad d_l^2 = \sum_{m=0}^{l} \left(\overline{C}_{lm}'^2 + \overline{S}_{lm}'^2\right) \tag{8}$$

where $\sigma_{C_{lm}}, \sigma_{S_{lm}}$ are the errors in the spherical harmonic coefficients of the geopotential obtained from the error covariance matrix and r_e is the radius of the Earth. Figure 5 shows a graph of the degree error variance of the geoid versus the degree variance of ζ. The errors in the geoid exceed the signal of ζ at about degree 10, thus justifying the decision to truncate the expansion of ζ at that degree. For comparison, the geoid degree error variance of a similar gravity field model computed without altimeter data is also shown. For this model, the errors in the geoid exceed the signal of ζ at degree 6. This explains why previous solutions for ζ which did not employ the simultaneous solution technique described here were truncated at degree 6 [*Douglas et al.*, 1984; *Tai and Wunsch*, 1984; *Engelis*, 1984].

CONCLUSIONS

A model of the sea surface topography of the oceans has been obtained using GEOSAT altimeter data in a simultaneous solution for the gravity field of the Earth and the orbits of satellites included in the solution. GEOSAT altimeter data and doppler tracking data were used in combination with tracking data from 13 other satellites and surface gravity data in order to obtain the solution. The geoid solution represents a significant improvement over previous solutions which did not employ altimeter data. The solution for ζ compares very well with models of ζ computed from oceanographic data, especially considering the large

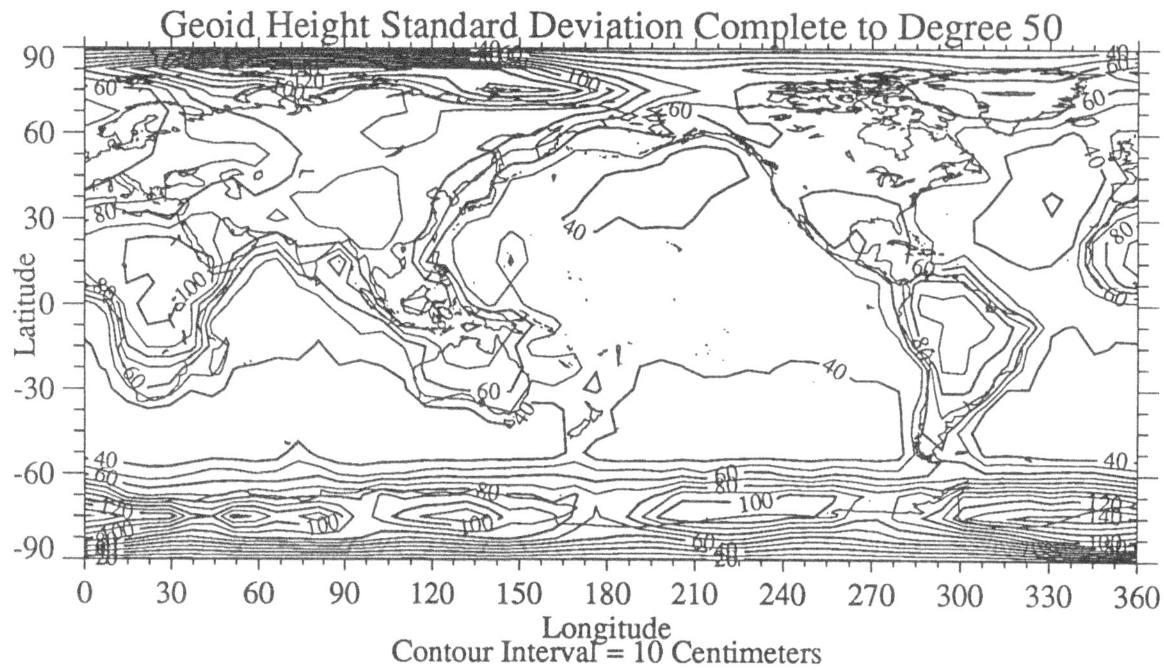

Fig. 3. Map of the geoid height standard deviation complete to degree 50.

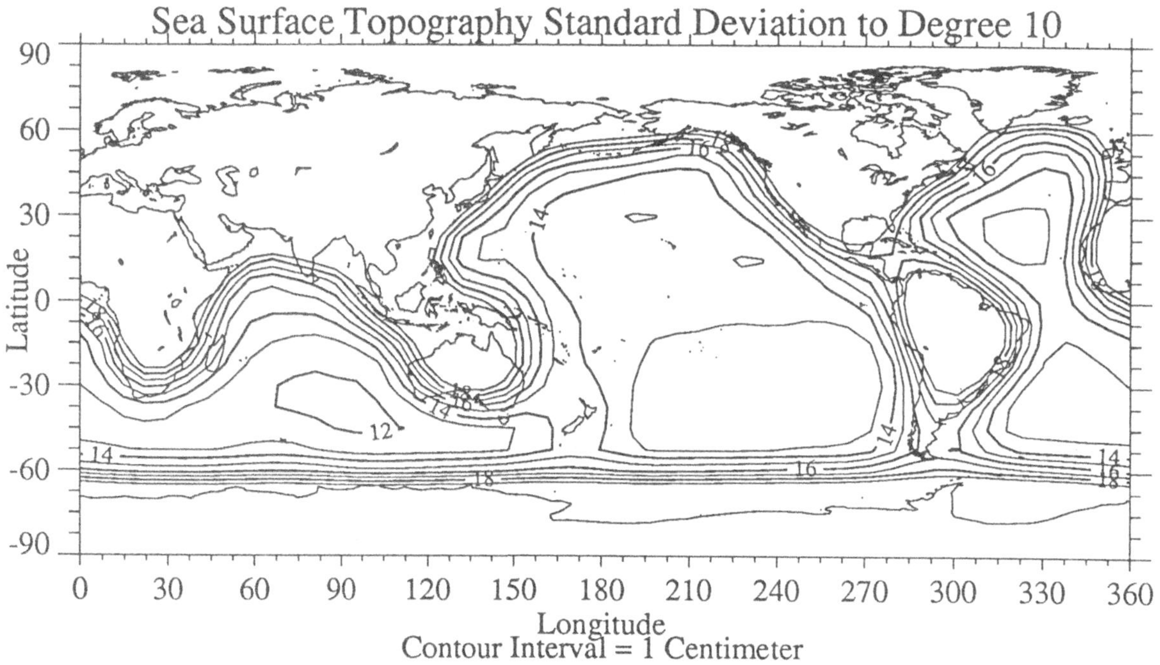

Fig. 4. Map of the standard deviation of ζ complete to degree 10

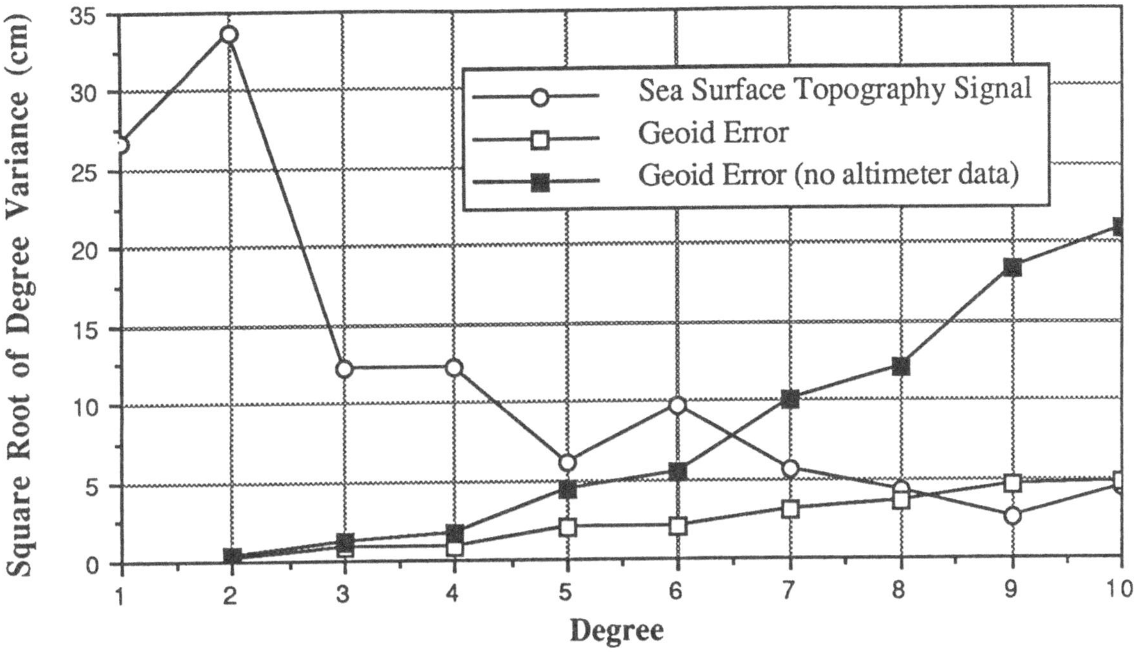

Fig. 5. Comparison of the geoid error and the signal of ζ versus degree

error sources present in both the oceanographic and altimetric techniques. The results shown here bode well for the success of the Topex/Poseidon mission [*Topex*, 1981] which will have more accurate altimeter measurements, orbit estimates, and atmospheric corrections. When combined with the hydrographic data collected during WOCE [*Nowlin*, 1987], the determination of the three dimensional velocity structure of the oceans is within reach.

Acknowledgment. The assistance of J. C. Ries and D. N. Yuan in the completion of this investigation is gratefully acknowledged. This work was supported by NASA Grant JPL-958122 and by the University of Texas System, Center for High Performance Computing.

REFERENCES

Coleman, R. (1980) On the Recovery of Ocean Dynamic Information from Satellite Altimetry, *Marine Geodesy*, Vol. 4, No. 4, pp 351-386.

Douglas, B. C., R. W. Agreen, and D. T. Sandwell (1984) Observing Global Ocean Circulation with Seasat Altimeter Data, *Marine Geodesy*, Vol. 8, No. 1-4, pp. 67-83.

Engelis, T., and R. H. Rapp (1984) Global Ocean Circulation Patterns Based on Seasat Altimeter Data and the GEM-L2 Gravity Field, *Marine Geophysical Research*, Vol. 7, pp. 55-67.

Engelis, T., and P. Knudsen (1989) Orbit Improvement and Determination of the Ocean Geoid and Topography from 17 Days of Seasat Data, *Manuscripta Geodetica*, Vol. 14, pp. 193-201.

Kaula, W. M. (1966) *Theory of Satellite Geodesy*, Blaisdell, Waltham, Mass.

Levitus, S. (1982) Climatological Atlas of the World Ocean, NOAA Professional Paper 13, Geophysical Fluid Dynamics Laboratory, Rockville, Maryland.

Marsh, J. G., F. J. Lerch, C. J. Koblinsky, S. M. Klosko, J. W. Robbins, R. G. Williamson, and G. B. Patel (1989a) Dynamic Sea Surface Topography, Gravity, and Improved Orbit Accuracies from the Direct Evaluation of SEASAT Altimetry Data, NASA Technical Memorandum 100735, Goddard Space Flight Center.

McConathy, D. R., and C. C. Kilgus (1987) The Navy Geosat Mission: An Overview, *Johns Hopkins APL Technical Digest*, 8(2).

Nerem, R. S. (1989) Determination of the General Ocean Circulation Using Satellite Altimetry From a Simultaneous Solution for the Earth's Gravity Field, Ph.D. Dissertation, The University of Texas at Austin.

Nowlin, W. D. (1987) World Ocean Circulation Experiment (WOCE), Status of US Planning for WOCE, *Bulletin of the American Meteorological Society*, Vol. 68, No. 12, pp. 1559-1565.

Pavlis, N. K. (1988) Modeling and Estimation of a Low Degree Geopotential Model From Terrestrial Gravity Data, Department of Geodetic Science and Surveying, The Ohio State University, Report No. 386.

Shum, C.-K. (1983) Altimeter Methods in Satellite Geodesy, Center for Space Research Report, CSR-83-2, The University of Texas at Austin.

Smith, J. C., J. C. Ries, C. K. Shum, B. E. Schutz, and B. D. Tapley (1988) Precision Orbit Determination for the Geosat Exact Repeat Mission, AIAA/AAS Astrodynamics Conference, Technical Papers, pp. 35-45.

Tai, C. -K., and C. Wunsch (1984) An Estimate of Global Absolute Dynamic Topography, *Journal of Physical Oceanography*, Vol. 14, pp. 457-463.

Tapley, B. D., R. S. Nerem, C. K. Shum, J. C. Ries, and D. N. Yuan (1988) Determination of the General Ocean Circulation From a Joint Gravity Field Solution, *Geophysical Research Letters*, Vol. 15, No. 10, pp. 1109-1112.

TOPEX (1981) Satellite Altimetric Measurements of the Ocean, Report of the TOPEX Science Working Group, NASA, JPL Doc. 400-111.

Wagner, C. A., and O. Colombo (1979) Gravitational Spectra from Direct Measurements, *Journal of Geophysical Research*, Vol. 84, pp. 4699-4712.

Wunsch, C. and E. M. Gaposchkin (1980) On using satellite altimetry to determine the general circulation of the ocean with application to geoid improvement, *Review of Geophysics and Space Physics*, Vol. 18, pp. 725-745

CHRISTOFFEL SYMBOLS AND THE LAPLACIAN IN DETAILED STUDIES OF THE EARTH´S GRAVITY FIELD WITH EMPHASIS ON TOPOGRAPHY AND ECCENTRICITY EFFECTS

Petr Holota
Research Institute of Geodesy, Topography and Cartography
250 66 Zdiby 98, Praha-východ, Czechoslovakia

INTRODUCTION

In geodesy the investigation of the Earth´s gravity field is particularly focused on its exterior part in a domain Ω given by the space outside the Earth or outside a model of the Earth (as, e.g., the telluroid). The boundary of this domain is rather complicated since it is as irregular as the physical surface of the Earth. As usual it is described by means of the height h above a reference ellipsoid of revolution. Suppose now that we have a system of geodetic coordinates H, B, L related to the reference ellipsoid of revolution. Here H is the geodetic height, B and L are geodetic latitude and longitude, respectively. Recall that the general point of geodetic coordinates H, B, L has the following rectangular Cartesian coordinates

$$x_1 = (N + H) \cos B \cos L \quad , \tag{1}$$

$$x_2 = (N + H) \cos B \sin L \quad , \tag{2}$$

$$x_3 = \left[(1 - e^2)N + H\right] \sin B \tag{3}$$

where

$$N = a(1 - e^2 \sin^2 B)^{-1/2} \tag{4}$$

is the principal radius of curvature of our reference ellipsoid of revolution in the direction of the prime vertical, a is the semiaxis of our reference ellipsoid and e is its numerical eccentricity. In addition, as is well known,

$$M = (1 - e^2)N^3 a^{-2} \tag{5}$$

is the principal radius of curvature of our reference ellipsoid of revolution in the direction of the meridian.

The physical surface of the Earth substantially differs from an ellipsoid of revolution. It may be closely approximated by the ellipsoid of revolution (or even by a sphere). However, the approximation is not valid for slopes and curvatures. The irregularity of the physical surface of the Earth causes difficulties in the solution of boundary value problems of physical geodesy related to the determination of the exterior gravity field and figure of the Earth from astrogeodetic and gravimetric data. The situation may be more convenient in a system of curvilinear coordinates such that the physical surface of the Earth (or of its model as above) is imbedded in one of the three families of coordinate surfaces. In this case, however, in the solution of boundary value problems of physical geodesy (as a role for the disturbing potential T) the structure of the Laplace operator of T is more complicated. In a sense its coefficient represent the topography of the physical surface of the Earth as we shall see later. This alternative between a boundary complication and a complication of coefficients of the partial differential equation governing the solution is often used in the solution of other boundary value problems of engineering and mathematical physics. In the next section we shall show how it can be materialized by means of a transformation of coordinates, cf. (Holota, 1989a,b).

TRANSFORMATION OF COORDINATES AND THE LAPLACIAN AS AN INVARIANT

Consider 3-dimensional Euclidean space E^3 and a mapping (transformation) F of our domain $\Omega \subset E^3$ given by the equations (1) - (4) with

$$H = Z + \omega(Z)h(B,L) \quad . \tag{6}$$

Here $\omega(Z)$ is a suitably chosen and twice continuously differentiable auxiliary function defined for $Z \in [0,\infty)$ such that

$$\omega(0) = 1 \quad , \quad \omega(\infty) = 0 \tag{7}$$

and

$$(1 - e^2)N + Z + \omega(Z)h(B,L) > 0 \tag{8}$$

for any values of the variables considered whereas $h(B,L)$ is the geodetic height which describes the boundary of Ω, as in the introduction. Clearly, the function $\omega(Z)$ gives the transformation F a higher degree of flexibility. Its closer specification enables, e.g., to simplify the structure of the boundary conditions, see (Holota, 1989a,b).

Now interpreting Z as the geodetic height again and supposing that

$$\frac{\partial H}{\partial Z} = 1 + \frac{d\omega}{dZ} > 0 \quad , \tag{9}$$

we immediately see that the transformation F is a one-to-one mapping between the domain Ω and the space outside our ellipsoid of revolution. Obviously, in the curvilinear coordinates Z, B, L the boundary of the domain Ω (i.e. the physical surface of the Earth or the telluroid) is defined by $Z = 0$ and, in the space of the coordinates Z, B, L, it coincides with the ellipsoid of revolution.

We are now in a position to calculate the fundamental metric tensor in the curvilinear coordinates Z, B, L . Putting (for purely technical reasons)

$$y_1 = Z \quad , \quad y_2 = B \quad , \quad y_3 = L \quad , \tag{10}$$

we immediately see that the Jacobian

$$J = |\partial x_i/\partial y_j| = (1 + \frac{d\omega}{dZ} h)(N + H)(M + H) \cos B \tag{11}$$

of the transformation F is positive in the domain $F^{-1}(\Omega)$ (apart from zero values for $B = -\pi/2$ and $\pi/2$). Consider first the components of the metric tensor

$$g_{ij}(y) = \frac{\partial x_k}{\partial y_i}\frac{\partial x_k}{\partial y_j} \tag{12}$$

in the coordinates y_i . In our particular case, going back to the original notation (10), we arrive at

$$g_{11} = (1 + \frac{d\omega}{dZ} h)^2 \quad , \tag{13}$$

$$g_{12} = g_{21} = (1 + \frac{d\omega}{dZ} h)\, \omega \frac{\partial h}{\partial B} \quad , \tag{14}$$

$$g_{13} = g_{31} = (1 + \frac{d\omega}{dZ} h)\, \omega \frac{\partial h}{\partial L} \quad , \tag{15}$$

$$g_{22} = (M + H)^2 + \omega^2 (\frac{\partial h}{\partial B})^2 \quad , \tag{16}$$

$$g_{23} = g_{32} = \omega^2 \frac{\partial h}{\partial B}\frac{\partial h}{\partial L} \quad , \tag{17}$$

$$g_{33} = (N + H)^2 \cos^2 B + \omega^2 (\frac{\partial h}{\partial L})^2 \tag{18}$$

with H given by (6). In addition

$$g = |g_{ij}| = J^2 \quad . \tag{19}$$

Similarly, denoting the cofactor of g_{ij} in the determinant g by G^{ij} and putting

$$g^{ij} = G^{ij}/g \tag{20}$$

for the components of the associated metric tensor, after an elementary calculation we obtain

$$g^{11} = (1 + \frac{d\omega}{dZ} h)^{-2} [1 + \omega^2 (M + H)^{-2} (\frac{\partial h}{\partial B})^2 + \\ + \omega^2 (N + H)^{-2} (\frac{\partial h}{\partial L})^2 \cos^{-2} B] \quad , \tag{21}$$

$$g^{12} = g^{21} = - (1 + \frac{d\omega}{dZ} h)^{-1} (M + H)^{-2}\, \omega \frac{\partial h}{\partial B} \quad , \tag{22}$$

$$g^{13} = g^{31} = - (1 + \frac{d\omega}{dZ} h)^{-1} (N + H)^{-2}\, \omega \frac{\partial h}{\partial L} \cos^{-2} B \quad , \tag{23}$$

$$g^{22} = (M + H)^{-2} \quad , \tag{24}$$

$$g^{23} = g^{32} = 0 \quad , \tag{25}$$

$$g^{33} = (N + H)^{-2} \cos^{-2} B \tag{26}$$

and recall, again, that H is given by (6).

We now have all the quantities which enable us to write immediately

$$\Delta T = g^{-1/2} \frac{\partial}{\partial y_i} (g^{1/2} g^{ij} \frac{\partial T}{\partial y_j}) = g^{ij} \frac{\partial^2 T}{\partial y_i \partial y_j} + g^{-1/2} \frac{\partial}{\partial y_i} (g^{1/2} g^{ij}) \frac{\partial T}{\partial y_j} \quad (27)$$

for the Laplace operator of T in terms of the curvilinear coordinates y_i, see (Sokolnikoff, 1971, eq. 92.11). However, we will prefer another expression for ΔT which has a closer relation to the general definition of ΔT. According to this definition ΔT is the second covariant derivative of T, or the first covariant derivative of the gradient of T, contracted with the associated metric tensor g^{ij} to form an invariant. Following Sokolnikoff (1971, eq. 92.8), we have

$$\Delta T = g^{ij} (\frac{\partial^2 T}{\partial y_i \partial y_j} - \Gamma_{ij}^{k} \frac{\partial T}{\partial y_k}) \quad (28)$$

where Γ_{ij}^{k} is the Christoffel symbol of the second kind

$$\Gamma_{ij}^{k} = g^{kl} [ij,l] \quad (29)$$

and

$$[ij,k] = \frac{1}{2} (\partial g_{jk}/\partial y_i + \partial g_{ik}/\partial y_j - \partial g_{ij}/\partial y_k) \quad (30)$$

is the Christoffel symbol of the first kind. In comparison with (27), formula (28) is more convenient if additional transformations of the coordinates have to be applied, see (Holota, 1989b). (Note. Within the general theory these transformations enable us to take into account perturbations coming, e.g., from the structure of the boundary conditions.)

CHRISTOFFEL SYMBOLS OF THE FIRST KIND

In this section, recalling the obvious symmetry, $[ij,k] = [ji,k]$, we give a full set of Christoffel symbols of the first kind in the coordinates y_i. Going back to the original notation (10), we obtain by direct calculation

$$[11,1] = (1 + \frac{d\omega}{dZ} h) \frac{d^2\omega}{dZ^2} h \quad , \quad (31)$$

$$[11,2] = \omega \frac{d^2\omega}{dZ^2} h \frac{\partial h}{\partial B} \quad , \quad (32)$$

$$[11,3] = \omega \frac{d^2\omega}{dZ^2} h \frac{\partial h}{\partial L} \quad , \quad (33)$$

$$[12,1] = (1 + \frac{d\omega}{dZ} h) \frac{d\omega}{dZ} \frac{\partial h}{\partial B} \quad , \quad (34)$$

$$[12,2] = (1 + \frac{d\omega}{dZ} h)(M + H) + \omega \frac{d\omega}{dZ} (\frac{\partial h}{\partial B})^2 \quad , \quad (35)$$

$$[12,3] = \omega \frac{d\omega}{dZ} \frac{\partial h}{\partial B} \frac{\partial h}{\partial L} \quad , \quad (36)$$

$$[13,1] = (1 + \frac{d\omega}{dZ} h) \frac{d\omega}{dZ} \frac{\partial h}{\partial L} \quad , \tag{37}$$

$$[13,2] = \omega \frac{d\omega}{dZ} \frac{\partial h}{\partial B} \frac{\partial h}{\partial L} \quad , \tag{38}$$

$$[13,3] = (1 + \frac{d\omega}{dZ} h)(N + H) \cos^2 B + \omega \frac{d\omega}{dZ} (\frac{\partial h}{\partial L})^2 \quad , \tag{39}$$

$$[22,1] = - (1 + \frac{d\omega}{dZ} h)[(M + H) - \omega \frac{\partial^2 h}{\partial B^2}] \quad , \tag{40}$$

$$[22,2] = 3e^2 a^{-2} N^2 M(M + H) \sin B \cos B + [(M + H) + \omega \frac{\partial^2 h}{\partial B^2}] \omega \frac{\partial h}{\partial B} \quad , \tag{41}$$

$$[22,3] = - [(M + H) - \omega \frac{\partial^2 h}{\partial B^2}] \omega \frac{\partial h}{\partial L} \quad , \tag{42}$$

$$[23,1] = (1 + \frac{d\omega}{dZ} h) \omega \frac{\partial^2 h}{\partial B \partial L} \quad , \tag{43}$$

$$[23,2] = (M + H) \omega \frac{\partial h}{\partial L} + \omega^2 \frac{\partial h}{\partial B} \frac{\partial^2 h}{\partial B \partial L} \quad , \tag{44}$$

$$[23,3] = - (N + H)^2 [1 - e^2 a^{-2} N^3 (N + H)^{-1} \cos^2 B] \sin B \cos B +$$

$$+ (N + H) \omega \frac{\partial h}{\partial B} \cos^2 B + \omega^2 \frac{\partial h}{\partial L} \frac{\partial^2 h}{\partial B \partial L} \quad , \tag{45}$$

$$[33,1] = - (1 + \frac{d\omega}{dZ} h)[(N + H) \cos^2 B - \omega \frac{\partial^2 h}{\partial L^2}] \quad , \tag{46}$$

$$[33,2] = (N + H)^2 [1 - e^2 a^{-2} N^3 (N + H)^{-1} \cos^2 B] \sin B \cos B -$$

$$- [(N + H) \cos^2 B - \omega \frac{\partial^2 h}{\partial L^2}] \omega \frac{\partial h}{\partial B} \quad , \tag{47}$$

$$[33,3] = [(N + H) \cos^2 B + \omega \frac{\partial^2 h}{\partial L^2}] \omega \frac{\partial h}{\partial L} \tag{48}$$

where $H = Z + \omega(Z) h(B,L)$, see (6).

CHRISTOFFEL SYMBOLS OF THE SECOND KIND

It is obvious that a relation of symmetry is also valid for the Christoffel symbols of the second kind. Indeed, we have

$$\Gamma^k_{ij} = \Gamma^k_{ji} \quad . \tag{49}$$

Now, as in the last section, a direct calculation yields

$$\Gamma^1_{11} = (1 + \frac{d\omega}{dZ} h)^{-1} \frac{d^2\omega}{dZ^2} h \quad , \tag{50}$$

$$\Gamma_{11}^{2} = 0 \quad , \tag{51}$$

$$\Gamma_{11}^{3} = 0 \quad , \tag{52}$$

$$\Gamma_{12}^{1} = - \left[\omega(M + H)^{-1} - (1 + \frac{d\omega}{dZ} h)^{-1} \frac{d\omega}{dZ}\right] \frac{\partial h}{\partial B} \quad , \tag{53}$$

$$\Gamma_{12}^{2} = (1 + \frac{d\omega}{dZ} h)(M + H)^{-1} \quad , \tag{54}$$

$$\Gamma_{12}^{3} = 0 \quad , \tag{55}$$

$$\Gamma_{13}^{1} = - \left[\omega(N + H)^{-1} - (1 + \frac{d\omega}{dZ} h)^{-1} \frac{d\omega}{dZ}\right] \frac{\partial h}{\partial L} \quad , \tag{56}$$

$$\Gamma_{13}^{2} = 0 \quad , \tag{57}$$

$$\Gamma_{13}^{3} = (1 + \frac{d\omega}{dZ} h)(N + H)^{-1} \quad , \tag{58}$$

$$\begin{aligned} \Gamma_{22}^{1} = &- (1 + \frac{d\omega}{dZ} h)^{-1}(M + H) - \\ &- (1 + \frac{d\omega}{dZ} h)^{-1}(M + H)^{-1}\left[3e^{2}a^{-2}N^{2}M \sin B \cos B + 2\,\omega \frac{\partial h}{\partial B}\right] \omega \frac{\partial h}{\partial B} + \\ &+ (1 + \frac{d\omega}{dZ} h)^{-1}\, \omega \frac{\partial^{2} h}{\partial B^{2}} \quad , \end{aligned} \tag{59}$$

$$\Gamma_{22}^{2} = (M + H)^{-1}\left[3e^{2}a^{-2}N^{2}M \sin B \cos B + 2\,\omega \frac{\partial h}{\partial B}\right] \quad , \tag{60}$$

$$\Gamma_{22}^{3} = 0 \quad , \tag{61}$$

$$\begin{aligned} \Gamma_{23}^{1} = &(1 + \frac{d\omega}{dZ} h)^{-1}\left[1 - e^{2}a^{-2}N^{3}(N + H)^{-1}\cos^{2}B\right] \omega \frac{\partial h}{\partial L} \operatorname{tg} B - \\ &- (1 + \frac{d\omega}{dZ} h)^{-1}\left[(M + H)^{-1} + (N + H)^{-1}\right] \omega \frac{\partial h}{\partial B} \frac{\partial h}{\partial L} + \\ &+ (1 + \frac{d\omega}{dZ} h)^{-1}\, \omega \frac{\partial^{2} h}{\partial B \partial L} \quad , \end{aligned} \tag{62}$$

$$\Gamma_{23}^{2} = (M + H)^{-1}\, \omega \frac{\partial h}{\partial L} \quad , \tag{63}$$

$$\Gamma_{23}^{3} = - \left[1 - e^{2}a^{-2}N^{3}(N + H)^{-1}\cos^{2}B\right] \operatorname{tg} B + (N + M)^{-1}\, \omega \frac{\partial h}{\partial B} \quad , \tag{64}$$

$$\Gamma^{1}_{33} = - (1 + \frac{d\omega}{dZ} h)^{-1}(N + H) \cos^2 B -$$

$$- (1 + \frac{d\omega}{dZ} h)^{-1}(M + H)^{-2}(N + H)^2[1 -$$

$$- e^2a^{-2}N^3(N + H)^{-1}\cos^2 B] \; \omega \frac{\partial h}{\partial B} \sin B \cos B -$$

$$- 2(1 + \frac{d\omega}{dZ} h)^{-1}(N + H)^{-1} \; \omega^2 (\frac{\partial h}{\partial L})^2 + (1 + \frac{d\omega}{dZ} h)^{-1} \; \omega \frac{\partial^2 h}{\partial L^2} \quad , \tag{65}$$

$$\Gamma^{2}_{33} = (M + H)^{-2}(N + H)^2[1 - e^2a^{-2}N^3(N + H)^{-1}\cos^2 B] \sin B \cos B \quad , \tag{66}$$

$$\Gamma^{3}_{33} = 2(N + H)^{-1} \; \omega \frac{\partial h}{\partial L} \tag{67}$$

where, again, $H = Z + \omega(Z)h(B,L)$, see (6).

LAPLACE OPERATOR - CONSTRUCTION

We are now in the position to write the Laplace operator of T in terms of the curvilinear coordinates Z, B, L . However, going back to formula (28), we still have to calculate

$$c_k = - g^{ij} \; \Gamma^{k}_{ij} \quad . \tag{68}$$

Recalling our notation (10), we have

$$c_1 = (1 + \frac{d\omega}{dZ} h)^{-1}[(M + H)^{-1} + (N + H)^{-1}] -$$

$$- (1 + \frac{d\omega}{dZ} h)^{-1} \; \omega \; [(M + H)^{-2} \frac{\partial^2 h}{\partial B^2} - (M + H)^{-2} \frac{\partial h}{\partial B} \operatorname{tg} B +$$

$$+ (N + H)^{-2} \frac{\partial^2 h}{\partial L^2} \cos^{-2} B] +$$

$$+ (1 + \frac{d\omega}{dZ} h)^{-1} \; \omega \; e^2a^{-2}N^2(M + H)^{-2}[3M(M + H)^{-1} -$$

$$- N(N + H)^{-1}] \frac{\partial h}{\partial B} \sin B \cos B +$$

$$+ 2(1 + \frac{d\omega}{dZ} h)^{-2} \; \omega \frac{d\omega}{dZ} [(M + H)^{-2}(\frac{\partial h}{\partial B})^2 + (N + H)^{-2}(\frac{\partial h}{\partial L})^2 \cos^{-2} B] -$$

$$- (1 + \frac{d\omega}{dZ} h)^{-3}[1 + \omega^2(M + H)^{-2}(\frac{\partial h}{\partial B})^2 +$$

$$+ \omega^2(N + H)^{-2}(\frac{\partial h}{\partial L})^2 \cos^{-2} B] \frac{d^2\omega}{dZ^2} h \quad , \tag{69}$$

$$c_2 = - (M + H)^{-2} \operatorname{tg} B -$$

$$- e^2a^{-2}N^2(M + H)^{-2}[3M(M + H)^{-1} - N(N + H)^{-1}] \sin B \cos B \quad , \tag{70}$$

$$c_3 = 0 \tag{71}$$

where $H = Z + \omega(Z)h(B,L)$ in view of (6).

Using these results, together with (21) - (26), we are able to write

$$\Delta T = g^{11}\frac{\partial^2 T}{\partial Z^2} + g^{22}\frac{\partial^2 T}{\partial B^2} + g^{33}\frac{\partial^2 T}{\partial L^2} + 2g^{12}\frac{\partial^2 T}{\partial Z\partial B} + 2g^{13}\frac{\partial^2 T}{\partial Z\partial L} + c_1\frac{\partial T}{\partial Z} + c_2\frac{\partial T}{\partial B} \quad . \tag{72}$$

Note that for $h = 0$ evidently $H = Z$ and

$$\Delta_e T = (\Delta T)_{h=0} = \alpha_1\frac{\partial T}{\partial Z} + \frac{\partial^2 T}{\partial Z^2} + \alpha_2\frac{\partial T}{\partial B} + \alpha_3\frac{\partial^2 T}{\partial B^2} + \alpha_4\frac{\partial^2 T}{\partial L^2} \quad , \tag{73}$$

with

$$\alpha_1 = (M + Z)^{-1} + (N + Z)^{-1} \quad , \tag{74}$$

$$\alpha_2 = -(M + Z)^{-2}\mathrm{tg}\, B - e^2a^{-2}N^2(M + Z)^{-2}[3M(M + Z)^{-1} - N(N + Z)^{-1}]\sin B\cos B \; , \tag{75}$$

$$\alpha_3 = (M + Z)^{-2} \quad , \tag{76}$$

$$\alpha_4 = (N + Z)^{-2}\cos^{-2}B \quad , \tag{77}$$

has the structure of the Laplace operator expressed in terms of Z, B, L taken for geodetic coordinates related to our ellipsoid of revolution, cf. (Holota, 1986, 1989b).

Similarly for $e = 0$ we have

$$\Delta_h T = (\Delta T)_{e=0} = (1 + \omega\frac{h}{r})^{-2}[\Delta_s T - \delta(T,h,\omega)] \tag{78}$$

where

$$r = a + Z \quad , \tag{79}$$

$$\Delta_s T = \frac{1}{r^2}(2r\frac{\partial T}{\partial r} + r^2\frac{\partial^2 T}{\partial r^2} - \frac{\partial T}{\partial B}\mathrm{tg}\, B + \frac{\partial^2 T}{\partial B^2} + \cos^{-2}B\frac{\partial^2 T}{\partial L^2}) \quad , \tag{80}$$

$$\delta(T,h,\omega) = A_1\frac{\partial T}{\partial r} + A_2\frac{\partial^2 T}{\partial r^2} + A_3\frac{1}{r}\frac{\partial^2 T}{\partial r\partial B} + A_4\frac{1}{r\cos B}\frac{\partial^2 T}{\partial r\partial L} \tag{81}$$

and A_i are topography-dependent coefficients given by the equations

$$A_1 = (1 + \frac{d\omega}{dr}h)^{-1}[2(\frac{d\omega}{dr} - \frac{\omega}{r})\frac{h}{r} + \frac{\omega}{r^2}\Delta_1 h] - 2(1 + \frac{d\omega}{dr}h)^{-2}\frac{\omega}{r^2}\frac{d\omega}{dr}|\mathrm{grad}_1 h|^2 + (1 + \frac{d\omega}{dr}h)^{-3}[(1 + \omega\frac{h}{r})^2 + (\frac{\omega}{r})^2|\mathrm{grad}_1 h|^2]\frac{d^2\omega}{dr^2}h \quad , \tag{82}$$

$$A_2 = (1 + \frac{d\omega}{dr}h)^{-2}[2(\frac{d\omega}{dr} - \frac{\omega}{r})h + ((\frac{d\omega}{dr})^2 - (\frac{\omega}{r})^2)h^2 - (\frac{\omega}{r})^2|\mathrm{grad}_1 h|^2] \quad , \tag{83}$$

$$A_3 = 2\left(1 + \frac{d\omega}{dr} h\right)^{-1} \frac{\omega}{r} \frac{\partial h}{\partial B} \quad , \tag{84}$$

$$A_4 = 2\left(1 + \frac{d\omega}{dr} h\right)^{-1} \frac{\omega}{r \cos B} \frac{\partial h}{\partial L} \tag{85}$$

with

$$|\mathrm{grad}_1 h|^2 = \left(\frac{\partial h}{\partial B} \right)^2 + \left(\frac{1}{\cos B} \right)^2 \left(\frac{\partial h}{\partial L} \right)^2 \tag{86}$$

and

$$\Delta_1 h = \left(\frac{1}{\cos B} \right) \frac{\partial}{\partial B} \left(\cos B \frac{\partial h}{\partial B} \right) + \left(\frac{1}{\cos B} \right)^2 \frac{\partial^2 h}{\partial L^2} \quad , \tag{87}$$

cf. (Holota, 1989a,b). Recall that $|\mathrm{grad}_1 h|^2$ and $\Delta_1 h$ are Beltrami´s first and second differential operators related to the unit sphere, respectively. (Note on this occasion that there is a misprint in (Holota, 1989a) in the coefficient A_1 .)

Following the analogy for $e = 0$, we first deduce that

$$(M + H)^{-1} = (N + H)^{-1}(1 + e^2 q) \quad , \quad q = a^{-2} N^3 (M + H)^{-1} \cos^2 B \quad . \tag{88}$$

Now

$$\Delta T = \left(1 + \frac{N - a}{r} + \omega \frac{h}{r} \right)^{-2} \left[\Delta_s T - \delta(T,h,\omega) - \delta_e(T,h,\omega) - \delta_h(T,h,\omega) \right] \tag{89}$$

where $\Delta_s T$ and $\delta(T,h,\omega)$ are given by (80) and (81), respectively, while

$$\begin{aligned} \delta_e(T,h,\omega) = & - \left(1 + \frac{d\omega}{dr} h\right)^{-1} \left[2(N - a) + e^2 q(N + H)\right] r^{-2} \frac{\partial T}{\partial r} - \\ & - \left(1 + \frac{d\omega}{dr} h\right)^{-2} \frac{N - a}{r} \left[2\left(1 + \omega \frac{h}{r}\right) + \frac{N - a}{r} \right] \frac{\partial^2 T}{\partial r^2} + \\ & + (2e^2 q + e^4 q^2)\left[1 + \frac{(1 + e^2 q)^2}{2 + e^2 q} \left(3 \frac{M}{N} - \frac{M + H}{N + H}\right)\right] r^{-2} \frac{\partial T}{\partial B} \operatorname{tg} B - \\ & - (2e^2 q + e^4 q^2) r^{-2} \frac{\partial^2 T}{\partial B^2} \quad , \end{aligned} \tag{90}$$

$$\delta_h(T,h,\omega) = (2e^2 q + e^4 q^2)\left(a_1 \frac{\partial T}{\partial r} + a_2 \frac{\partial^2 T}{\partial r^2} + a_3 \frac{1}{r} \frac{\partial^2 T}{\partial r \partial B} \right) \tag{91}$$

and

$$\begin{aligned} a_1 = & - \left(1 + \frac{d\omega}{dr} h\right)^{-1} \left[1 + \frac{(1 + e^2 q)^2}{2 + e^2 q} \left(3 \frac{M}{N} - \frac{M + H}{N + H}\right)\right] \frac{\omega}{r^2} \frac{\partial h}{\partial B} \operatorname{tg} B + \\ & + \left(1 + \frac{d\omega}{dr} h\right)^{-1} \frac{\omega}{r^2} \frac{\partial^2 h}{\partial B^2} - 2\left(1 + \frac{d\omega}{dr} h\right)^{-2} \frac{\omega}{r^2} \frac{d\omega}{dr} \left(\frac{\partial h}{\partial B} \right)^2 + \\ & + \left(1 + \frac{d\omega}{dr} h\right)^{-3} \left[2 \frac{N - a}{r} \left(1 + \omega \frac{h}{r}\right) + \left(\frac{N - a}{r}\right)^2 + \right. \\ & \left. + \left(\frac{\omega}{r}\right)^2 \left(\frac{\partial h}{\partial B} \right)^2 \right] \frac{d^2 \omega}{dr^2} h \quad , \end{aligned} \tag{92}$$

$$a_2 = - \left(1 + \frac{d\omega}{dr} h\right)^{-2} \left(\frac{\omega}{r}\right)^2 \left(\frac{\partial h}{\partial B} \right)^2 \quad , \tag{93}$$

$$a_3 = 2(1 + \frac{d\omega}{dr} h)^{-1} \; \frac{\omega}{r} \frac{\partial h}{\partial B} \quad . \tag{94}$$

Note that for $h = 0$ and the coefficients at the individual derivatives of T expressed with an accuracy up to the degree of e^2 we have

$$\delta_e(T,0,\omega) \doteq - e^2 \frac{a}{r} \left[\frac{1}{r} \frac{\partial T}{\partial r} + \frac{\partial^2 T}{\partial r^2} \sin^2 B - \right.$$

$$\left. - \frac{2}{r^2} \left(2 \frac{\partial T}{\partial B} \operatorname{tg} B - \frac{\partial^2 T}{\partial B^2}\right) \cos^2 B \right] \tag{95}$$

which is in full agreement with eq. (2.23) in (Holota, 1986) as well as eq. (4.38) in (Holota, 1989b).

CONCLUSION

The disturbing potential T is a harmonic function in the domain Ω. In consequence T satisfies the Laplace equation

$$\Delta T = 0 \quad \text{for} \quad Z > 0 \tag{96}$$

which results in

$$\Delta_s T = \delta(T,h,\omega) + \delta_e(T,h,\omega) + \delta_h(T,h,\omega) \tag{97}$$

for $r = a + Z > a$ in view of (89) and (8). Thus, in the coordinates Z, B, L, the terms δ, δ_e and δ_h on the right-hand side of (97) express the change of the spherical structure of the Laplace equation caused by the topography and a global flattening of the boundary of Ω. The term δ_e and especially the term δ_h also express an interrelation between the topography and eccentricity effects.

The auxiliary function ω is not fully specified in this paper. If we are interested in the solution of boundary value problems in the domain Ω, the function ω can be additionally chosen so as to simplify the structure of the respective boundary conditions, see (Holota, 1989a,b).

Finally our aim is to solve (97) under a given boundary condition for T by means of the method of successive approximations, provided that δ, δ_e and δ_h can be taken for small perturbations. It is obvious that this approach makes it possible to take advantage of the separability of the Laplace operator in spherical coordinates and to apply a spherical apparatus (as, e.g., the Green´s - Stokes´ function) at each iteration step, see (Holota, 1989a). Therefore, in future papers we will further develop and generalize the technique discussed in (Holota, 1989a,b), using the results achieved here.

Clearly a generalization of this technique for perturbations related to other than a spherical configuration is substantialy connected with a proper choice of coordinates since, in general, the threedimensional Laplacian is not separable for any system of coordinates, see (Grafarend, 1988).

Acknowledgment. Thanks are expressed to Prof. Dr. Erik W. Grafarend, Stuttgart University, for his interest and valuable comments.

REFERENCES

Grafarend, E.W. (1988). The Geometry of the Earth´s Surface and the Corresponding Function Space of the Terrestrial Gravitational Field. Festschrift zum sechzigsten Geburtstag von R. Sigl. Deutsche Geodätische Kommission, Bayerische Akad. d. Wiss., München, Heft Nr. 287, pp. 76-94.

Holota, P. (1986). Oblique Derivative Boundary Condition and the Ellipticity of the Solution Domain in the Theory of the Geodetic Boundary Value Problem. Proc. Int. Symp. on Figure and Dynamics of the Earth, Moon, and Planets, Prague, 1986. Astron. Inst. of the Czechoslovak Acad. of Sci. and Research Inst. of Geod., Topogr. and Cartogr., Prague, 1987, Part I, pp. 385-400.

Holota, P. (1989a). Laplacian Versus Topography in the Solution of the Molodensky Problem by Means of Successive Approximations. Festschrift to Torben Krarup, E. Kejlsø, K. Poder and C.C. Tscherning (eds.), Geodaetisk Institut, København, Meddelelse No. 58, pp. 213-227.

Holota, P. (1989b). Higher Order Theories in the Solution of Boundary Value Problems of Physical Geodesy by Means of Successive Approximations, invited paper. Proc. 2nd Hotine-Marussi Symp. on Math. Geodesy, Pisa (in print).

Sokolnikoff, I.S. (1971). Tensor Analysis, Theory and Applications to Geometry and Mechanics of Continua. Nauka Publishers, Moscow (in Russian).

CORRECTIONS FOR ATMOSPHERIC REFRACTION IN PRECISE LEVELING

Li Yanxing
Geodetic Brigade, State Seismological Bureau
Tianjin, P.R. of China

ABSTRACT

The error of atmospheric refraction is one of the main systematic errors in precise leveling. Through many years of tests and studies, the author has selected a model of temperature more applicable to practice and derived a strict formula for calculating the atmospheric refraction in leveling. Satisfactory result has been obtained in the test levelings, thus the foundation has been laid for the practical application in the future.

INTRODUCTION

The influence of atmospheric refraction exists in leveling. The experimental studies and the theoretical analysis show that the effect of atmospheric refraction on leveling is systematic in nature and it is one of the main systematic errors in precise leveling. The effect of refraction on leveling running in the mountain areas is significant. Therefore, studing the law of the refraction error is quite neccessary for making corrections to the leveling.

I. THE MODEL FOR VERTICAL TEMPERATURE DISTRIBUTION IN THE ATMOSPHERE NEAR THE EARTH SURFACE

In leveling, the height of sight line is rather near the earth surface, generally less than 3m. The temperature gradient in this layer of atmosphere is subjected to the influences from many factors. It varies all the time and the law of variation is rather complicated. At present, it is still very difficult to solve the problem of corrections for atmospheric refraction in precise leveling. In order to solve this problem, many scholars both in China and abroad have performed quite a lot of experimental studies, some temperature models have been established and formulas for corrections of refraction been derived. These formulas reveal to different extents the approximate law of refraction in leveling and have taken a certain effect in weakening the influence of refraction in leveling. But the main problem in these formulas is that the corrections are not accurate enough. For some data, the accuracy is not improved after correction but decreased instead. So these formulas have not been widely used. For exploring the law of atmospheric refraction, the author has been engaged in the experimental studies for many years and discovered that the corrected results for refraction mainly depend on the selection of temperature model and the method of calculation for the model selected. In addition, it is also related to the observation method for temperature.

Based on the experimental studies, the following temperature models can be selected:

$$t = a + bz^{c}$$

$$t = a + bz^{c} + dz$$

$$t = a + b \ln z$$

$$t = a + b \ln z + cz$$

where, z denotes the height above the ground; t the temperature at the height of z; a, b, c, d the coefficients to be determined. Based on the same observed data of leveling and temperature, the refraction corrections related to different models are different. The differences are very large, sometimes even up to 0.02mm for elevation difference per meter.

Model 1 and 2 are of the exponential type and it is very difficult to calculate a, b, c, d — the coefficients to be determined. The following methods can be used for accurate calculation.

1. The generalized estimation iterative method.
2. The damping least squares iterative method.
3. The method of direct calculation for c.
4. The least squares method.

The derived corrections for refraction are different, if different methods of calculation are used. Sometimes, the difference can be 0.03mm for elevation difference per meter.

The model 1 and 2 can be solved by all of the 4 methods and model 3 and 4 be solved by the least squares method. The four models and four methods can form 10 calculation schemes. Based on the experiment data and through a great deal of calculations, the temperature models and solution methods are sieved. Two criteria are proposed by the author for testing the effects of various models and solution methods.

1. The accuracy of the observed data should be improved after correction for refraction is made, that is the standard error for elevation difference per kilometer should be smaller after correction than that before correction.
2. After the observed elevation difference over the whole leveling line is corrected for refraction, the discreteness of elevation differences in 5 levelings should be decreased as compaired with that before the correction is made, that is the standard error for the observed elevation difference over the leveling line after correction should be smaller than that before correction.

Based on these two criteria, the different temperature models and solution methods are tested (details about the tests are not included here) and the following conclusions are made:

1. The "best" temperature model is

$$t = a + bz^c \tag{1}$$

2. The "best" solution method is the third method.

II. Deriving the Formula for Refraction Corrections

In temperature model (1), there are three coefficients to be determined — a, b and c. So long as 3 or more sets of observed data (t_i, z_i) are obtained, the three coefficients can be derived then. Since the model is an exponential function, it is very difficult to calculate simultaneously the three coefficiens a, b, and c. Therefore, some scholars let the c-value be fixed and equal to -1/3; others use the graphical method to obtain the c-value. After the c-value is determined, the values a and b can be derived with the least squares method. But the c-value changes all the time with the meteorological and the territorial conditions. So it is not suitable to suppose that the c-value is fixed and the graphical method is neither convenient nor accurate. By many years of researches, a strict method to directly calculate the c-value has been devised by the author. It is discribed as follows,

If the values of z are taken as $z_1=0.5$, $z_2=1.5$ and $z_3=2.5$ and the corresponding atmospheric temperatures are t_1, t_2 and t_3, then based on equation (1), we have,

$$t_1 = a + 0.5^c = a + 0.5^c \times 1 \tag{2}$$
$$t_2 = a + 1.5^c = a + 0.5^c \times 3^c \tag{3}$$
$$t_3 = a + 2.5^c = a + 0.5^c \times 5^c \tag{4}$$

from (3)−(2) and (4)−(2), we get

$$t_2 - t_1 = 0.5^c \times (3^c - 1) \tag{5}$$
$$t_3 - t_1 = 0.5^c \times (5^c - 1) \tag{6}$$

from (5) / (6) with $t_3 - t_1 \neq 0$, we have

$$(t_1 - t_2) \times 5^c + (t_3 - t_1) \times 3^c + t_2 - t_3 = 0 \tag{7}$$

let

$$y = f(c) = (t_1 - t_2) \times 5^c + (t_3 - t_1) \times 3^c + t_2 - t_3 = 0$$

One can know from the experiments that the domain of c-value is approximately [−3, 3]. The calculation is made successively one point by one point in the domain of c-value from −3 to 3 with a small increment Δc (for instance $\Delta c=0.02$) to obtain $y=f(c_i)$. Given a minor value ε close to zero (e.g $\varepsilon=0.01$), when $c_i=c_0$, we have $|y_0| = |f(c_0)| \leqslant \varepsilon$, then c_0 is taken as the approximate value of c. After the c-value is derived, the values a and b can be solved by the least square method, that is the following formulas

$$3a+[z^c]b-[t]=0$$
$$[z^c]a+[(z^c)^2]b-[tz^c]=0$$

Sometimes, the c-value fitted to $|y|\leqslant \varepsilon$ is not unique, generally, there is a set of c-values. Based on each c_i, the corresponding a_i and b_i can be derived. Substitute a set of a_i, b_i and c_i into the following equation

$$V_i=[V_j\ V_j]=[(a_i+b_i z_j^{c_i}-t_j)^2]$$

and the corresponding V_i is obtained . In this set of V-values, the values a, b, c corresponding to the minimum V-value are the required best probable values.

The above-mentioned process can be expressed by the following equations

$$\begin{aligned}&(t_1-t_2)\times 5^c+(t_3-t_1)\times 3^c+t_2-t_3=0\\&3a+[z^c]b-[t]=0\\&[z^c]a+[(z^c)^2]b-[tz^c]=0\\&[(a+bz^c-t)^2]=\min.\end{aligned}\tag{8}$$

Similarly, if the values of z are taken as $z_1=0.5$, $z_2=1.5$, $z_3=2.0$, $z_4=2.5$m and the corresponding atmospheric temperatures are t_1, t_2 t_3 and t_4, the coefficients a, b, and c can be derived by equations (9).

$$\begin{aligned}&(t_1-t_3)\times 5^c+(t_4-t_2)\times 4^c+(t_3-t_1)\times 3^c+t_2-t_4=0\\&4a+[z^c]b-[t]=0\\&[z^c]a+[(z^c)^2]b-[tz^c]=0\\&[(a+bz^c-t)^2]=\min.\end{aligned}\tag{9}$$

If the values of z are taken as $z_1=0.5$, $z_2=1.0$, $z_3=1.5$, $z_4=2.0$ $z_5=2.5$m and the corresponding atmospheric temperatures are t_1, t_2 t_3 t_4 and t_5, the coefficients a, b, and c can be derived by equations (10).

$$\begin{aligned}&(3t_1+t_2-t_3-3t_4)\times 5^c+(t_1-t_2-3t_3+3t_5)\times 4^c+\\&(3t_4+t_5-t_1-3t_2)\times 3^c+(3t_3+t_4-3t_1-t_5)\times 2^c+\\&3t_2+t_3-t_4-3t_5=0\\&5a+[z^c]b-[t]=0\\&[z^c]a+[(z^c)^2]b-[tz^c]=0\\&[(a+bz^c-t)^2]=\min.\end{aligned}\tag{10}$$

Now, the first equation in (10) can be proved as follows. Based on equation (1), we get

$$t_1=a+b\times 0.5^c\times 1 \quad (11)$$
$$t_2=a+b\times 0.5^c\times 2^c \quad (12)$$
$$t_3=a+b\times 0.5^c\times 3^c \quad (13)$$
$$t_4=a+b\times 0.5^c\times 4^c \quad (14)$$
$$t_5=a+b\times 0.5^c\times 5^c \quad (15)$$

Taking arbitrarily 3 equations from the above 5 equations each time, one can derive a set of a, b, c values and there are 10 combinations in these 5 equations, 1,2,3; 1,2,4; 1,2,5; 1,3,4; 1,3,5; 1,4,5; 2,3,4; 2,3,5; 2,4,5 and 3,4,5.

Taking 1, 2, 3, we have

$$(t_3-t_2)\times 1^c+(t_1-t_3)\times 2^c+(t_2-t_1)\times 3^c=0 \quad (16)$$

Taking 1, 2, 4, we have

$$(t_4-t_2)\times 1^c+(t_1-t_4)\times 2^c+(t_2-t_1)\times 4^c=0 \quad (17)$$

.
.
.

Taking 3, 4, 5, we have

$$(t_5-t_4)\times 3^c+(t_3-t_5)\times 4^c+(t_4-t_3)\times 5^c=0 \quad (25)$$

Adding up each side of the above 10 equations (from (16) to (25)), we have

$$(3t_1+t_2-3t_4-t_3)\times 5^c+(t_1-t_2-3t_3+3t_5)\times 4^c+$$
$$(3t_4+t_5-t_1-3t_2)\times 3^c+(3t_3+t_4-3t_1-t_5)\times 2^c+$$
$$3t_2+t_3-t_4-3t_5=0$$

The sight line in leveling can be expressed by the following equation

$$y=z_0+\int_0^x \beta\, dx \quad (26)$$

where, z_0 denotes the height of the sight line at the leveling station; β the angle between the tangent line and the horizontal plane at any point along the sight line.

Suppose i_0 is the incidence angle of sight line at the leveling station; i the incidence angle at any point along the sight line; $\bar{n}$ the atmospheric refraction index at the leveling station; n the refraction index at any point along the sight line; $\eta_0=1.0002923$ denotes the standard atmospheric refraction index; P_0 (=760/mmHg) represents the standard atmospheric pressure and P the average atmospheric pressure along the leveling segment during observation; h the elevation difference for each leveling station; s the length of sight distance; t_0 the temperature at the leveling station; t the temperature at any point along the sight line; α the angle between the earth surface and the horizontal plane; z the height above the earth surface at any point along the sight line. Based on the research in literature [1], we have

$$\beta = \alpha + i - \frac{\pi}{2}$$

$$i_0 = \frac{\pi}{2} - \alpha$$

$$i = \arcsin\left(\frac{\bar{n}}{n}\cos\alpha\right)$$

$$\bar{n} = 1 + \frac{n_0 - 1}{1 + \gamma t_0}\frac{p}{p_0}$$

$$n = 1 + \frac{n_0 - 1}{1 + \gamma t}\frac{p}{p_0}$$

$$\alpha = \operatorname{arctg}\frac{h}{2s}$$

$$z = z_0 - x\operatorname{tg}\alpha$$

Expanding i in series at $x = 0$

$$i = i_0 + \dot{i}_0 x + \frac{1}{2}\ddot{i}_0 x^2 + \cdots\cdots$$

The formula (26) can be written as

$$y = z_0 + \int_0^x (\dot{i}_0 x + \frac{1}{2}\ddot{i}_0 x^2 + \cdots\cdots)\,dx$$

Integrating (26), we have

$$y = z_0 + \frac{1}{2}\dot{i}_0 x^2 + \frac{1}{6}\ddot{i}_0 x^3 + \cdots\cdots$$

Suppose V_{fore}, V_{back} and V represent respectively the errors caused by the atmospheric refraction in the readings (on the fore and back staffs) and in the elevation difference at the leveling station, we have

$$V_{fore} = y_x - y_0 = +\frac{1}{2}\dot{i}_0 x^2 + \frac{1}{6}\ddot{i}_0 x^3 + \cdots\cdots$$

$$V_{back} = y_{-x} - y_0 = +\frac{1}{2}\dot{i}_0 x^2 - \frac{1}{6}\ddot{i}_0 x^3 + \cdots\cdots$$

$$V = V_{back} - V_{fore} = -\frac{1}{3}\ddot{i}_0 x^3 - \cdots\cdots$$

Suppose δ is the refraction correction for the elevation difference at a leveling station. Substituting x by s, we have

$$\delta = -V = \frac{1}{3}\ddot{i}_0 S^3 + \cdots\cdots$$

Omitting derivatives of the third order and higher order, we have

$$\delta = \frac{1}{3} \dot{i}_o S^3 + \cdots\cdots \tag{27}$$

Deriving (27) for i, we get

$$\dot{i}_o = ebcz_o^{c-1} tg\,\alpha \{ (c-1) z_o^{-1} + (bcz_o^{c-1}+1) \times [e(2+ctg^2\alpha) - 2\gamma (1+\gamma t_o)^{-1}] \} \tag{28}$$

where,

$$e = \bar{n}^{-1} (n_o - 1)(1+\gamma t_o)^{-2} p\, p_o^{-1} \gamma$$

Substituting equation (28) into (27), we have

$$\delta = \frac{1}{6} S^2 hebcz_o^{c-1} \{ (c-1) z_o^{-1} + 2(bcz_o^{c-1}+1) \times [e(1+2S^2h^{-2}) - \gamma (1+\gamma t_o)^{-1}] \} \tag{29}$$

Since

$$|2(bcz_o^{c-1}+1)[e(1+2S^2h^{-2}) - \gamma(1+\gamma t_o)^{-1}]| << |(c-1) z_o^{-1}|$$

the equation (29) can be simplified as

$$\delta = \frac{1}{6} S^2 hebc (c-1) z_o^{c-2} \tag{30}$$

Equations (29) and (30) are just the formulas for calculating the correction of refraction in the elevation difference at a leveling station.

III. EXPERIMENTAL RESULTS AND DISCUSSIONS

In order to study the atmospheric refraction in leveling, a testing line has been selected in the western suburbs of Beijing. The testing line was 20.7km long which was divided into 7 leveling segments with different lengths and slopes. The total elevation difference of the line was about 288m. The line was surveyed for 5 times with double-run levelings. The lengths of sight line were less than 50m for 3 levelings, less than 35m for one leveling and less than 20m for another. Temperature data were observed for each setup during the levelings by a automatic telethemometer which has 5 sensors distributed at the different heights of 0.5, 1.0, 1.5, 2.0, and 2.5m above the earth surface. An average elevation difference of double-run leveling can be obtained for each time of double-run leveling along each segment. Total of 35 average elevation differences can be obtained for 5 times of double-run leveling along 7 segments. Thus the leveling standard errors for one kilometer and for the whole testing line can be calculated for both the original observed value and the one of refraction-corrected. The standard error in leveling for one kilometer can be calculated by the following formula

$$\mu = \pm \{ [\sum_{i=1}^{7} \sum_{j=1}^{5} (V_{ij} V_{ij}) \div S_i] \div (35-7) \}^{1/2} \tag{31}$$

where, V_{ij} is the difference between the jth observation in the ith leveling segment and the average value from 5 observations; S_i denotes the length of the ith segment.

For the whole line, one can calculate the average value and the standard error with and without corrections. The formulas for calculation are as follows

$$\bar{h} = (\sum_{i=1}^{5} h_i) \div 5 \tag{32}$$

$$m = \pm [\sum_{i=1}^{5} (V_{ij} V_{ij}) \div (5-1)]^{1/2} \tag{33}$$

where, h_i is the average elevation difference for the ith double-run leveling over the whole line, $V_i = \bar{h} - h_i$. The calculated results are shown in the following table

	m mm	M mm	[△] mm	△m mm	△h mm
Results without refraction correction	0.478	3.215	-5.625	8.213	3.598
Results with refraction correction	0.465	2.193	+0.924	5.532	0.264

[△] is the sum of discrepancies for 35 leveling segments.
△m is the difference between the maximum and the minimum elevation differences over the whole leveling line.
m is the standard error per kilometer calculated from the mean elevation differences of 35 leveling segments.
M is the standard error of elevation difference over the whole leveling line.
△h is the difference over the whole leveling line between the average elevation differences for three times double-run levelings with long sight line and the one for two times double-run levelings with short sight line.

It can be seen from the above table, though the experiments were preliminary, the automatic telethermometer used was not very accurate and the devices for experiments were not perfect, the experiments were successful. The standard errors both for one kilometer and for the whole leveling line were reduced and the difference between the maximum and the minimum elevation difference for the whole leveling line was obviously decreased when the refraction correction was added. As a result, the accuracy of the leveling data were improved.

While the leveling survey is carrying out, if the temperatures at different heights in the atmosphere below 3m are observed directly by an automatic telethemometer and making corrections based on the above-metioned formulas to the leveling data, it may be a right way for resolving the refraction-correction problem in leveling. If the automatic telethemometer and total device for temperature observation can be developed even more accurate and perfect, the effect of refraction correction in leveling would be further improved. This device for temperature observation is placed on a bike and only one person is needed. Making refraction corrections to the leveling data can basically eliminate the influence of atmonspheric refraction and improve the accuracy of the leveling data. At the same time, the length of sight line in leveling can be lengthed and the amounts of observations carried out in a.m. and p.m. respectively are not neccessary to be strictly equal. Thus the velocity and the efficiency in leveling can be raised.

REFERENCE

1) Li Yanxing, An accurate formula for refraction corrections in precise leveling (in Chinese), ACTA GEODETICA et CARTOGRAPHICA SINICA, Vol.13, No.2, 1984.

Symposium 104 Author Index

GPSR Compliance

The European Union's (EU) General Product Safety Regulation (GPSR) is a set of rules that requires consumer products to be safe and our obligations to ensure this.

If you have any concerns about our products, you can contact us on ProductSafety@springernature.com

In case Publisher is established outside the EU, the EU authorized representative is:

Springer Nature Customer Service Center GmbH
Europaplatz 3
69115 Heidelberg, Germany

Batch number: 10855247

Printed by Printforce, the Netherlands